Future Power Generation Applications, Challenges, and Solutions

Edited by Mohamed Zellagui

Published in London, United Kingdom

Future Power Generation Applications, Challenges, and Solutions
http://dx.doi.org/10.5772/intechopen.1008582
Edited by Mohamed Zellagui

Contributors
Aakanksha Bedi, Agus Krisnowo, Bei Gou, David Bogen, Gani Soehadi, Ignatius Subagjo, I Made Astawa, Irawan Santoso, Iwan Sudrajat, J. Ramprabhakar, Jaizuluddin Mahmud, Mohamad Dahsyat, Mohamed Zellagui, Mohammed Gronfula, Oo Abdul Rosyid, Qing Zhao, R. Anand, Susalit Setya Wibowo, Syaeful Karim, Ugay Sugarmansyah, Yustiar Gunawan

First published in London, United Kingdom, 2026 by IntechOpen
IntechOpen is the global imprint of INTECHOPEN LIMITED, registered in England and Wales, registration number: 11086078, 167-169 Great Portland Street, London, W1W 5PF, United Kingdom

For EU product safety concerns: IN TECH d.o.o., Prolaz Marije Krucifikse Kozulić 3, 51000 Rijeka, Croatia, info@intechopen.com or visit our website at intechopen.com.

British Library Cataloguing-in-Publication Data
A catalogue record for this book is available from the British Library

Future Power Generation Applications, Challenges, and Solutions
Edited by Mohamed Zellagui
p. cm.
Print ISBN 978-1-83635-686-8
Online ISBN 978-1-83635-685-1
eBook (PDF) ISBN 978-1-83635-687-5

Meet the editor

Associate Professor Mohamed Zellagui received the B. Eng. (Hons.) and M. Sc. degrees in Electrical Power Engineering from the University of Constantine 1, Algeria, in 2007 and 2010, respectively, and the Ph.D. degree in Power Systems from the University of Batna 2, Algeria, in 2014. He is currently an Associate Professor in the Department of Electromechanical Engineering at the University of Batna 2, Faculty of Technology. He has authored or co-authored more than 50 refereed journal articles and conference papers, as well as nine book chapters and five edited books. He is a member of the CIGRE and a senior member of the IEEE. His research interests include power systems, renewable energy, energy efficiency, optimization algorithms, and smart grids.

Contents

Preface

Future Power Generation Applications, Challenges, and Solutions synthesizes essential knowledge of power systems and renewable energy challenges and solutions.

This book helps professionals develop the technical and strategic skills required to support reliable, efficient, and sustainable electricity production. As the energy landscape evolves, professionals must stay up to date on innovations, environmental requirements, and performance optimization techniques. Power Generation training courses empower individuals to ensure a stable energy supply through smart operations and forward-thinking practices.

This book supports readers from initial understanding to design and implementation. Readers can build an understanding of generation technologies from distributed generation in power systems to consumers, while also looking at the future smart grid.

Providing in-depth analysis of the biggest challenges currently facing the industry. This book presents various challenges and solutions, enabling researchers, industry engineers, and students to develop solutions for the power systems of the future. It focuses on power generation, transmission, and distribution systems, providing skills and coding approaches for designing solutions to industry challenges.

Dr. Mohamed Zellagui
Department of Electromechanical Engineering,
Faculty of Technology,
University of Batna 2,
Batna, Fesdis, Algeria

Section 1

Introduction

Chapter 1

Introductory Chapter: Overview of AC Power Generation and Energy Management

Mohamed Zellagui

1. Introduction

A reciprocating engine and an alternator serve as the foundation of power generators, which are compact, independent power plants. Depending on the amount of energy required, the engine and alternator are typically housed in a single enclosure that can be as large as a tractor-trailer or as small as a suitcase [1]. There are generators with engines that can run on a variety of fuels, including natural gas, diesel, gasoline, and biofuels.

There are numerous businesses and applications for larger power generators. They can function as backup power or primary power [2]. For instance, military installations situated in regions lacking a dependable electrical grid frequently rely on on-base generators to meet all their electricity requirements.

Standby generators are installed to supply electricity in the event of a power loss in buildings where it is unacceptable to lose power at any time. Buildings that require standby generators include hospitals and data centers [3].

The use of renewable energy sources to generate electricity has gained popularity in recent years. Electricity is rapidly being produced using renewable energy sources like geothermal, solar, wind, and hydro. Because they naturally replenish and do not eventually run out, these energy sources are regarded as renewable.

Renewable energy sources are becoming more affordable compared to conventional fossil fuels and nuclear energy, in addition to their advantages for the environment.

2. AC power generation system components

The production of alternating current (AC) power is a complex process that requires the cooperation of several key elements to convert energy from various sources into electricity [3–6]:

- Power generator,
- Prime mover,
- Control system,

- Power electronics,
- Voltage regulator.

3. Energy management and power grid integration

Despite the difficulties involved, switching to renewable energy offers several opportunities for innovation in electrical grid integration and energy management [5–7]:

- Smart distribution grids,
- Micro-grids and nano-grids,
- Battery technologies,
- Renewable hybrid systems,
- Inverter technologies,
- Real-time control.

4. Conclusion

The contemporary electrical distribution system is continuously evolving, mainly due to the increasing integration of renewable energy sources and the associated challenges related to energy management, power losses, and voltage issues. The main objective in power network optimization is to minimize total real power loss while maintaining acceptable voltage profiles.

All things considered, power generation is a complicated and important facet of contemporary civilization. To guarantee that the energy sources utilized are secure, dependable, and long-term sustainable, rigorous management and planning are necessary in the AC distribution and transmission power systems.

Author details

Mohamed Zellagui
Department of Electromechanical Engineering, University of Batna 2, Batna, Algeria

*Address all correspondence to: m.zellagui@univ-batna2.dz

References

[1] Rahman MM, Dadon SH, He M, Giesselmann M, Hasan MM. An overview of power system flexibility: High renewable energy penetration scenarios. Energies. 2024;**17**(24):6393. DOI: 10.3390/en17246393

[2] Zellagui M, Belbachir N, El-Sehiemy RA. Solving the optimal power flow problem in power systems using the Mountain Gazelle Algorithm. Engineering Proceedings. 2023;**56**(1):176. DOI: 10.3390/ASEC2023-16269

[3] Rodríguez-Benítez OM, Aqui-Tapia JA, Ortega-Velázquez I, Espinosa-Pérez G. Current source topologies for photovoltaic applications: An overview. Electronics. 2022;**11**(18):2953. DOI: 10.3390/electronics11182953

[4] Fujisawa N, Uchiyama T, Iio S. Review on the performance of crossflow hydraulic turbine for local power generation: Fundamentals and applications. Next Energy. 2025;**8**:100346. DOI: 10.1016/j.nxener.2025.100346

[5] Kumar MD, Biswas N, Sinha MP, Sarkar S, Manna NK. A critical review of photovoltaic cell integrated solar chimney: Sustainability and power generation. Solar Energy. 2024;**284**:113032. DOI: 10.1016/j.solener.2024.113032

[6] Zellagui M, Hassan HA, Kraimia MN. DFIG wind turbine under unbalanced power system conditions using adaptive fuzzy virtual inertia controller. Energetika. 2016;**65**(1):1–13. DOI: 10.6001/energetika.v65i1.3971

[7] Arunkumar T, Wilson HM, Pandit TP, Lee SJ. Electrical power generation and utilization in advanced desalination systems. Renewable and Sustainable Energy Reviews. 2025;**210**:115211. DOI: 10.1016/j.rser.2024.115211

Section 2

Electrical Power Systems

Chapter 2

Blockchain-Based VPP for Futuristic Power Generation and Management

Aakanksha Bedi, J. Ramprabhakar and R. Anand

Abstract

The global energy landscape is undergoing a fundamental transformation from centralized generation to a decentralized sustainable energy system. This paradigm shift is primarily driven by the rapid proliferation of distributed energy resources (DERs) such as solar photovoltaics, wind turbines, battery energy storage, responsive real-time demand response, and energy management. The virtual power plant (VPP) concept has emerged as an advanced framework to integrate, coordinate, and optimize these diverse DERs, enabling them to operate as a unified, flexible power entity. However, conventional centralized power plants face many challenges in terms of scalability, trust, security, and transparent coordination among all operating units. Blockchain technology, with its decentralized ledger, immutability, and smart contract capabilities, offers a transformative solution to these limitations. When such revolutionary technology is integrated with VPPs, it facilitates transparent peer-to-peer (P2P) energy trading, Nega-watt trading, and kilo watt trading. Hence, such a blockchain-based VPP (BVPP) framework establishes a decentralized energy management system that ensures trustless coordination, data integrity, and real-time optimization of renewable assets. This chapter explores the architecture, operational mechanisms, and potential applications of BVPP in the context of futuristic power systems. It discusses how blockchain enhances energy democratization, fosters prosumer participation, and optimizes transactive energy models. Furthermore, it addresses critical technical, economic, and regulatory challenges such as scalability, interoperability, and governance and proposes viable solutions for large-scale BVPP deployment.

Keywords: virtual power plant, renewable energy, blockchain technology, energy management, energy trading

1. Introduction

1.1 Overview of the energy landscape transformation

For decades, the global power system has been built around a centralized generation model, or centralized power plants (CPPs), where large power plants such as coal, hydro, or gas produce electricity at a few specific locations. This electricity is then transmitted over long distances through high-voltage transmission lines and distributed to consumers through a radial network. Thus, in such a system, power always flowed one way: from generation units to transmission systems to distribution stations and then finally to consumers. However, such a model is struggling to cope with the rapid adoption of renewable energy sources, especially solar and wind, because such resources are distributed, intermittent, and variable in nature. Unlike centralized plants, they can be located at homes, factories, or community microgrids, making the grid no longer unidirectional. Solar panels and wind turbines introduce bi-directional or sometimes multi-directional power flows. Moreover, consumers, now called prosumers, can generate their own green electricity and also inject unutilized power into the grid. Hence, traditional power systems, designed for one-way energy flow, face voltage fluctuations, reverse power flow issues, frequency instability, and difficulties in load balancing when accommodating these dynamic resources [1].

Moreover, the centralized power system cannot efficiently handle thousands of small, dispersed distributed energy resources (DERs) due to limited communications, coordination, and data management capabilities. Furthermore, the centralized power system uses a top-down approach, but to manage DERs, we need a bottom-up approach. **Figure 1** explains the problem in the existing centralized power system. This transformation highlights the urgent need for new management systems, such as virtual power plants (VPPs), which can aggregate, optimize, and coordinate DERs in real time. These decentralized systems represent the next evolutionary step toward a smart, flexible, scalable, and sustainable energy network capable of achieving high renewable energy penetration while maintaining grid stability.

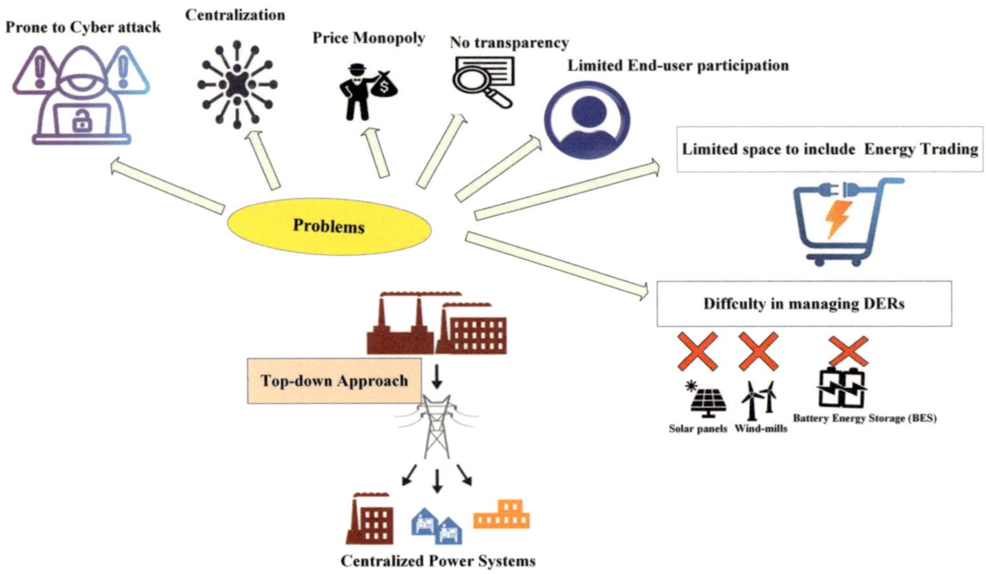

Figure 1.
Traditional centralized power system and its challenges in the current scenario.

1.2 Virtual power plant

A VPP is like a smart control energy management system, which can be thought of as a cloud of power that can smartly control energy generation, provide bi-directional power, and enable information flow. The foundation of a VPP is based on strong and advanced communication technologies. Thus, when a normal grid is equipped with many sensors, advanced communication devices, and is able to sustain bi-directional power flow, it is called a smart grid, and the VPP becomes an integral part of effectively managing these smart grids and DERs. In this way, a VPP can be thought of as the Internet of Energy (IoE), which is a cloud-based software that combines not only DERs and smart grids but also weather forecasting systems, energy market regulators, energy storage devices, microgrids, smart metering devices, schedulable or non-schedulable loads, demand response (DR), and most importantly, prosumers [2]. As a VPP is not location-specific and can operate virtually, it allows maximum utilization of locally generated energy without the need for a new generation unit. It also optimizes locally generated power from renewable energy sources and eliminates long-distance transmission. Moreover, for a greener and sustainable future, E-mobility will be a pioneering solution, and for this, a VPP will be the only solution for reducing grid burden by incorporating electric vehicles (EVs) into the vehicle-to-grid (V2G) mode. To be more precise, a VPP functions as a digital twin of the physical energy ecosystem, capable of orchestrating thousands of distributed assets into a single, dispatchable, and market-ready entity.

1.3 Blockchain technology

Blockchain is a decentralized, distributed, and immutable ledger technology (DLT) where data is stored across multiple computers known as nodes or participants instead of being kept in a single centralized database. Each of these nodes maintains a complete or partial copy of the blockchain ledger, meaning they all store the same set of verified transactions. All these transactions are bundled together in a block once the transactions are verified by the blockchain network. When any participant or node initiates a transaction, the transaction is broadcast to all nodes. All the other nodes validate the transactions, meaning they check for authenticity or digital signatures. Once validated, a consensus mechanism determines which node will get a chance to add these transactions into a new block. The new block is then cryptographically linked to the previous block through its hash, ensuring immutability, meaning once information is recorded, it cannot be altered without altering subsequent blocks. Once the block is added, all nodes update their copies of the blockchain to reflect this newly added block. This makes blockchain highly secure and trustworthy. Another core feature of blockchain is smart contracts, which are self-executing programs stored on the blockchain that automatically perform actions when predefined conditions are met. These smart contracts completely eliminate manual intervention and reduce human error [3].

In the energy sector, blockchain relevance is very impactful. It establishes trust among distributed participants such as DERs, consumers, prosumers, and utility operators by providing a transparent, verifiable record of every energy transaction. The immutability of blockchain ensures accurate metering and billing, while smart contracts automate settlement between various VPP entities. Overall, blockchain introduces transparency, security, and disintermediation into traditionally CPPs.

1.4 Motivation for integrating blockchain and VPP

The VPPs can work without blockchain but suffer from some critical limitations. Traditional VPPs rely on centralized databases that are vulnerable to data manipulation, single points of failure, and a lack of trust among stakeholders. Without blockchain, P2P energy trading remains opaque, settlements are delayed, and cyber-security risks increase due to reliance on third-party intermediaries. Managing thousands of DERs without a distributed verification system leads to coordination inefficiencies, scalability bottlenecks, and auditability challenges. Blockchain directly addresses these issues through cryptographically secured distributed ledgers, automated smart contracts, and consensus-based validation. This creates a trusted digital backbone for transactive energy markets, where both centralized utilities and distributed participants operate under a unified, tamper-proof system [4].

The integration of blockchain technology with VPPs represents a transformative step in the evolution of power systems, bridging the gap between traditional centralized utilities and decentralized renewable energy networks. The main point is that blockchain does not replace the conventional grid; rather, it enhances its operational intelligence and market participation. By enabling real-time, verifiable energy transactions, blockchain ensures that every kilowatt-hour generated, stored, or traded is accurately recorded and settled. This means centralized operators can confidently integrate distributed assets such as solar rooftops, community batteries, or microgrids without losing visibility or control. It provides granular traceability of power flows, better demand forecasting through transparent data exchange, and automated settlement systems that reduce administrative overheads and billing disputes. Ultimately, integrating blockchain with VPPs ensures operational trust, real-time optimization, improved cyber-security, and new revenue channels, driving a collaborative energy ecosystem where centralized and decentralized systems coexist seamlessly for more resilient, efficient, and data-driven smart power grids or microgrids.

2. Fundamentals of VPPs

2.1 VPP architecture and components

The VPPs represent the next evolutionary stage of power system management, designed to handle the increasing penetration of DERs and the complexity of bi-directional or even multi-directional power flows. Unlike traditional centralized systems, where energy moves uni directionally from generation to consumption, a VPP adopts a bottom-up decentralized approach, integrating multiple small-scale generators, storage units, and flexible loads into coordinated, intelligent, and responsive networks [5]. The architecture of a VPP can be understood with four key components, which are:

2.1.1 Renewable generation units

These units create a heterogeneous and distributed generation portfolio, capable of delivering clean, green energy locally while reducing transmission losses and dependency on fossil-based plants. Such units are:-

- **Solar PVs:** Such systems serve as the backbone of decentralized clean energy generation. They are variable in nature and highly dependent on solar irradiance and weather conditions, requiring real-time monitoring and prediction. The drawback of such systems is that they cannot generate electricity during night time.

- **Wind turbines:** Such systems generate power based on wind speed and direction. The output power is less than solar but can provide output during night time well.

- **Micro-turbines:** Small-scale gas or biogas turbines add dispatchable flexibility to the system. Unlike solar and wind, they can quickly ramp up or down, making them ideal for balancing fluctuations in renewable energy output.

- **Microgrids:** A VPP can be thought of as clusters of microgrids, which are localized energy networks integrating different renewable energy sources. They can operate in grid-connected or islanded mode and are crucial to maintaining grid stability and reliability.

2.1.2 Storage units

Battery energy storage systems (BESs) play a crucial role in ensuring the reliability and continuity of the VPPs. They act as both energy buffers and stabilizers, storing excess renewable energy during low-demand or high-generation periods and releasing it when generation drops or demand spikes. Modern BES units, integrated through advanced power electronic converters, can respond within milliseconds, providing essential services like frequency regulation, voltage control, and peak shaving. Their interoperability with EVs (V2G) further enhances flexibility and resilience, turning EV fleets into mobile energy reservoirs.

2.1.3 Controllable loads and DR

Controllable loads refer to demand assets that can adjust their consumption dynamically based on grid signals or economic incentives. Through the DR mechanism, non-critical loads such as HVAC systems, electric heaters, or heavy industrial motors can be scheduled or curtailed to maintain system balance during peak hours. VPPs leverage Internet of Things (IoT) sensors, smart meters, and automated control systems to monitor and modulate these loads in real time, aligning consumption patterns with available renewable generation.

2.1.4 VPP operator

At the heart of VPP lies the VPP operator, or VPPO, which is nothing but a cloud-based software platform that integrates data from all distributed components and optimizes overall system performance. The VPPO performs forecasting, scheduling, economic dispatch, continuously evaluates grid conditions, renewable generation, and also keeps track of market participation using advanced analytics, artificial intelligence, and communication technologies. Thus, the VPPO always tries to achieve optimal energy balance and cost efficiency while maintaining system reliability.

2.2 VPP classification

VPPs are classified into two types, namely commercial VPP (CVPP) and technical VPP (TVPP). Together, they integrate economic optimization with technical stability in decentralized energy systems.

2.2.1 Commercial VPP

A CVPP emphasizes market-driven, commercial operations. It focuses on aggregating diverse DERs across wide geographies to participate in energy markets. The business objectives of CVPPs are to maximize revenue, optimize trading strategies, and align DER operations with price signals rather than grid services. Thus, CVPPs turn distributed assets into scalable commercial portfolios, bridging DERs with energy markets and enabling value creation beyond mere generation.

2.2.2 Technical VPP

A TVPP emphasizes operational, grid-services, and system engineering functions. It typically aggregates DERs within defined geographic or network zones and dynamically balances local generation, consumption, and storage to deliver grid-centric services such as frequency regulation, voltage support, loss minimization, reactive power compensation, congestion relief, and other ancillary services. Thus, TVPPs are a tool for maintaining reliability and operational efficiency in a high-DER environment, supporting the transition from one-way to bi- or multi-directional power flows. They treat DERs as controllable grid assets, not just market commodities.

Therefore, a well-designed VPP, which embeds both CVPP and TVPP, is essential, enabling revenue generation via CVPP while ensuring grid services and system robustness via TVPP.

2.3 Key function of VPP

2.3.1 Forecasting and optimization

Forecasting in VPP involves the precise prediction of generation from renewable sources and demand behavior across the aggregated network, while optimization follows forecasting by determining resource dispatch strategies that maximize energy and economic efficiency. Accurate prediction of generation and demand helps VPP to create realistic schedule bids and dispatch plans. Once forecasts are available, the optimization layer schedules and dispatches an aggregated DER portfolio. The ultimate goal is to minimize penalties from deviations, maximize revenue, and reduce uncertainty on the grid.

2.3.2 Market participation

A VPP must engage with energy and ancillary service markets. This means bidding into wholesale day-ahead markets, intra-day markets, reverse markets, and possibly local flexibility markets. A VPP plays a very important role in adjusting bids whenever actual generation deviates from the forecasted quantity by functioning as a market actor, enhancing flexibility and DER participation as economic assets, and enabling the monetization of DR, local generation, and charging–discharging patterns of EVs or BESs.

2.3.3 Grid support services

Beyond trading, VPPs provide ancillary services and grid support services, which are extremely critical for system stability in high DER networks. These include frequency regulation (achieved by adjusting dispatchable assets and controllable storage devices within milliseconds), voltage support (by dynamic reactive power compensation using inverter-based resources), ramping services, congestion management, reactive power support, and black start assistance through microgrid-islanding capability, to name a few. Recent studies show that VPPs coordinate active and reactive power dispatch to meet both economic and network operational objectives.

3. Blockchain technology for decentralized energy management

Blockchain technology is increasingly recognized as a transformative tool for decentralized energy management, particularly in advancing VPP and accelerating progress toward net-zero energy grids. The following synthesis offers a technical yet accessible overview of the core concepts, architectures, and use cases relevant to blockchain integration in the energy sector:

3.1 Core concept relevant to blockchain integration

3.1.1 Distributed ledger technology

DLT establishes a shared, immutable record of transactions, serving as the "single source of truth" for all participants in decentralized energy systems. This ensures transaction transparency and traceability, regardless of entity boundaries, allowing utilities, regulators, and prosumers to seamlessly collaborate and audit energy trades and system operations.

3.1.2 Smart contracts

Smart contracts are programmable, self-executing agreements residing on the blockchain. In energy contexts, these contracts autonomously execute trades, payments, and grid balancing once predefined conditions are met, like price thresholds, demand-response events, etc. Thus, smart contracts help in minimizing manual intervention and improving operational efficiency.

3.1.3 Consensus mechanisms

Blockchain systems require consensus protocols to validate transactions and secure the network. In a decentralized system like blockchain, there is no central authority to verify or approve transactions. Therefore, nodes must collectively decide which transactions are valid, what order they should follow to add these transactions into blocks, and which block should come first. This dilemma can be solved using consensus algorithms. Different blockchains use different consensus algorithms, and by using these consensus mechanisms, all participating nodes agree on the same ledger state and allow only verified, valid transactions to be added to the chain. The consensus also makes sure that no single participant can control or manipulate the network. There are many different consensus algorithms, such as:

Proof of work (PoW): Used by the Bitcoin network and Ethereum 1.0, offers robust security but is energy-intensive, often unsuitable for sustainability-focused energy systems.

Proof of stake (PoS) and pure proof of stake (PPoS): PoS is used by Ethereum 2.0, while PPoS is used by the Algorand blockchain. Both of these consensus mechanisms drastically reduce energy consumption, relying on verifiers to verify the blocks. It is important to note that PPoS is much more superior to PoS in terms of transaction finality and scalability, which is paramount for energy sector applications where efficiency is crucial.

Proof of authority (PoA), practical Byzantine fault tolerance (PBFT), and delegated proof of stake (DPoS): Enable fast, low-energy consensus by restricting block validation to trusted entities, such as utilities regulators, but somewhat hinder the concept of true decentralization.

3.2 Blockchain architectures in power systems

3.2.1 Permissioned versus permissionless blockchain

Permissioned blockchain restricts participation to authorized entities such as utilities, regulators, and market operators, supporting compliance, privacy, and performance requirements in VPP operation. These models provide enhanced control and regulatory oversight, essential for critical infrastructure, while permissionless blockchain is fully decentralized and open, making it less suitable for tightly regulated energy networks due to issues with control, privacy, and scalability.

3.2.2 Layer 2 solutions

To facilitate high-frequency energy trading and real-time settlements in VPPs, Layer 2 scalability solutions such as side-chains, rollups, and state channels are employed to offload transactions from the main chain. These Layer 2 solutions are protocols built on top of a primary blockchain (Layer 1) to overcome scalability and cost limitations inherent in the base layer. This can be achieved by processing the majority of the transactions off-chain and only periodically anchoring the summarized or validated data back to the main chain (Layer 1). This allows for a large volume of micro-transactions, such as energy trades between dozens or hundreds of distributed resources, to occur instantly and without clogging the main blockchain, which is crucial for maintaining the responsiveness required in real-time grid operations. By reducing the number of transactions that need to be recorded on Layer1, operational costs, such as gas fees in blockchain, will be minimized, making the economic model of BVPP's energy trading scalable.

3.3 Key use cases in the VPP context

3.3.1 Energy trading

Blockchain enables direct, auditable transactions between participants, bypassing intermediaries. Prosumers can sell excess renewable energy either directly to the grid by enabling kilowatt trading (KW) or directly send it to near by neighbors, enabling peer-to-peer (P2P) trading, or promote users to save their energy in terms of negawatt (NW), creating transparent marketplaces and unlocking new business models in distributed energy.

3.3.2 Immutable metering and billing

Blockchain ensures the integrity of smart meter data by making readings tamper-proof and verifiable for consumers, prosumers, and auditors. This capability guarantees fair billing and transparent reconciliation, which are vital in a multi-party energy system.

3.3.3 Renewable energy certificate (REC) tracking

Using blockchain, RECs verifying the origin of consumed energy are tracked transparently and securely. This immutability and traceability help avoid double-spending and fraud, thereby supporting de-carbonization initiatives and simplifying the burden for utilities and end users.

3.3.4 Blockchain-based tokenization in energy

Tokenization involves converting measurable energy assets such as KW, NW, or P2P energy exchange into programmable digital tokens managed via blockchain smart contracts. These tokens carry embedded logic for tracking ownership and settlement issues in real time. In smart meters, tokenization transforms consumption and generation data into immutable, tradable tokenized units that are instantly accessible to authorized parties, enabling tracking and secure exchange of data. Tokenization can also be used in managing DERs in real time, as it can directly convert electricity into tokens (generated by rooftop solar or domestic batteries), which are easy to track and trade. Tokenization can also be used in VPP by enabling regular accounting and compensation of VPP participants and DR contributors. Moreover, tokenized energy not only handles consumption rights but can also be utilized in renewable attributes such as carbon offsets, RECs, incentivizing sustainability, and helping in the net-zero goal [6].

4. Transformative impact for sustainable energy: The blockchain-integrated VPP model

By integrating blockchain with VPPs (BVPPs), an advanced energy management architecture can be formed that merges traditional VPP operations with decentralized blockchain infrastructure to achieve trustless automation, secure energy transactions, and transparent data governance. In such a model, the blockchain layer acts as a decentralized control, replacing the conventional VPP control system that effectively coordinates various DERs, ESs, and DR mechanisms [7].

4.1 BVPP architecture and data flow

The BVPP framework represents an advanced energy management architecture that merges traditional VPP operation with decentralized blockchain-based control within a VPP to achieve trustless, automatic, secure energy transactions and transparent data governance. In such an infrastructure, the blockchain layer acts as a key decentralized control and verification entity, in which all the participating entities of the smart grids, such as prosumers, consumers, aggregators, VPPs, and the grid, are registered through smart contracts. This registration process involves cryptographically verifiable pseudo-identities assigned to each node by a trusted authority (TA), ensuring anonymity and preventing key escrow (centralized key control may compromise data security) vulnerabilities. Such a mechanism can be thought of as autonomous digital agreements stored on the blockchain. Once data is entered into the blockchain, it is end-to-end secure and immutable, but what if something happens to the data before entering the blockchain? Since smart meters, sensors, and IoT devices are the main data points that continuously record real-time parameters such as generation profiles, demand profiles, voltage, frequency, and so on; moreover such systems are hotspots for attackers to steal private data, hence encryption becomes essential for such systems. Advanced data encryption (AES), which is a symmetric key cryptography method suitable for bulk data encryption, can be used for this purpose. To provide confidentiality and integrity, AES can be further authenticated through elliptical curve cryptography (ECC)–based public key mechanisms. In such a scenario, the AES session key can be securely exchanged and digitally signed using ECC-based public-private key pairs. **Figure 2** shows the detailed representation of BVPP. Once validated, encrypted data packets are written into the blockchain ledger via consensus mechanisms such as PPoS, PoS, or PoA, depending on whether BVPP operates in permissionless or permissioned mode. Both of these blockchain networks, permissioned or permissionless, store every validated transaction in a block, and all these blocks are cryptographically linked to each other using SHA-256, producing a fixed-length digest that uniquely represents the data, forming an immutable chain of blocks that prevents data tampering and ensures the same copy of data is available across all nodes.

4.2 Decentralized VPP operation using smart contract

Smart contracts automate the aggregation and dispatch process traditionally handled by centralized control centers. Based on weather-forecasted data, market price signals, and real-time grid frequency, smart contracts automatically execute

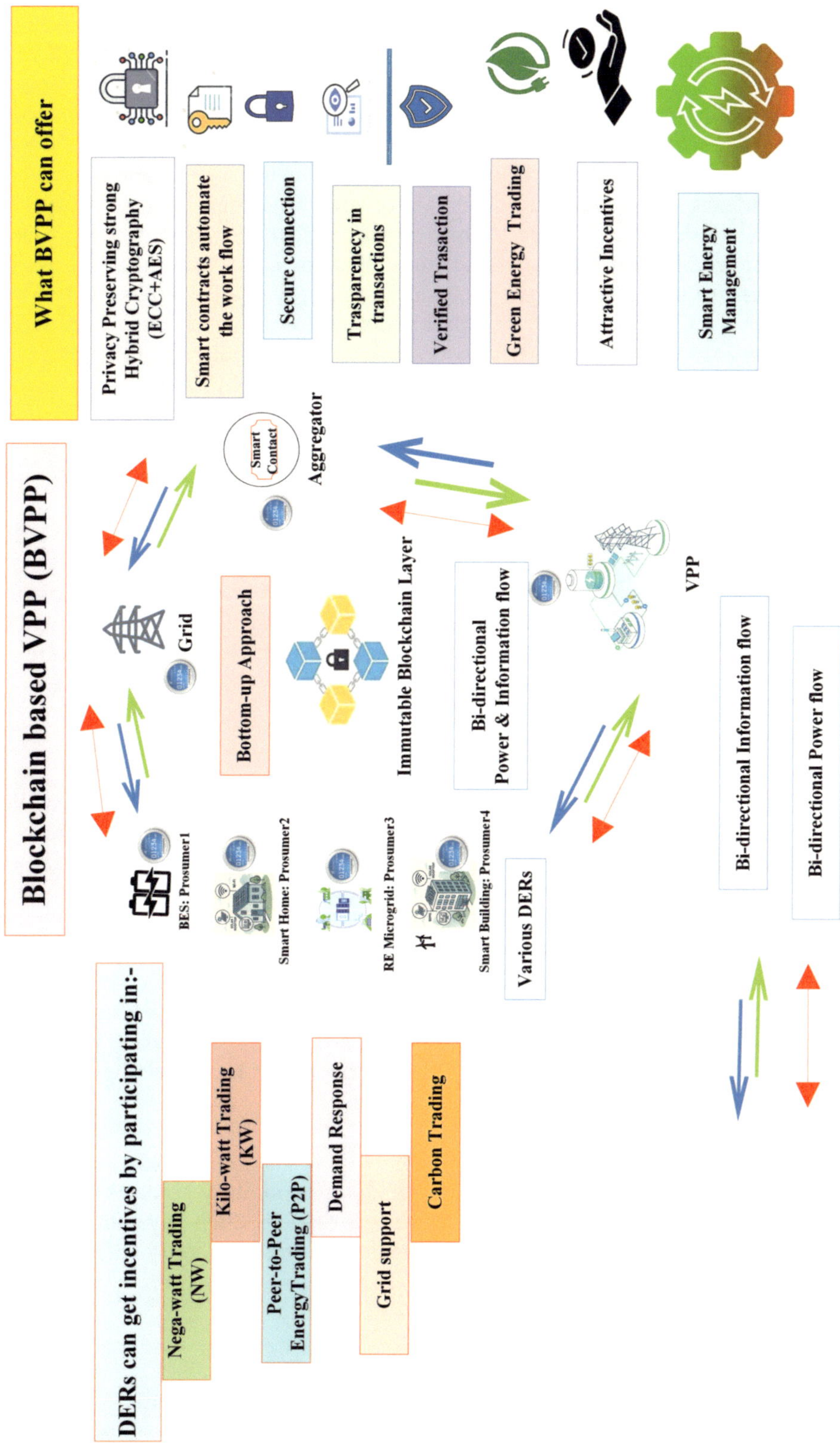

Figure 2.
Blockchain-based virtual power plant (BVPP) for dynamic real-time energy market.

the logic inbuilt into them, such as aggregating DER outputs from distributed prosumers, dispatching stored or generated energy during peak hours (KW trading), sending it to nearby (P2P trading), or saving energy in terms of kilowatts (KW trading) to support the grid and receive attractive incentives. Such operations occur without human intervention, as smart contracts are self-executing programs that activate when pre-defined conditions or thresholds are met. For example, if the grid frequency drops below a critical value or market price surpasses a preset threshold, the smart contract triggers battery discharge or prosumer feed-in from connected DERs to stabilize the grid. This decentralized mechanism allows real-time market clearing, enhanced liquidity within local energy markets, and minimizes transaction costs compared to traditional aggregator-led markets[8]. Hence, collectively, these features make the BVPP a secure, transparent, and decentralized framework capable of real-time energy optimization, cyber-secure communication, and fully automated market participation for achieving net-zero energy grids (NZEGs).

5. Applications and benefits of BVPP for future power generation

A BVPP combines VPP with blockchain primitives such as an immutable ledger, smart contracts, tokens, and transparency. This, together, enables trustless coordination, automatic settlement, market-clearing, and verifiable verification, allowing microgrid clusters and prosumers to trade energy, participate in ancillary markets, and jointly support the grid during peak times, opening new revenue streams.

5.1 Autonomous microgrids in BVPP

The BVPP framework allows microgrids to operate semi-independently while being part of a larger, aggregated resource pool. Every microgrid participant is authenticated and is free to settle transactions directly with peers using smart contracts, promoting flexible and dynamic local energy balance. Such a system improves grid stability, clearly detects congestion, and accordingly provides congestion management and other short. Usually, grid congestion occurs when electricity demand exceeds transmission or distribution capacity, causing bottlenecks and introducing instabilities in the overall system. A BVPP can mitigate congestion by enabling decentralized coordination of DERs, facilitating local energy generation and consumption, optimizing power flow, and incentivizing DR.

5.2 Optimized ancillary service provision

BVPP platform groups DERs to provide grid-scale ancillary services such as frequency regulation, voltage support, or spinning reserve. Such a system enables the pooling of DERs' capacity, meaning small residential batteries can become market participants.

5.3 Democratization of the energy systems

In the BVPP framework, power shifts to prosumers, meaning energy trading reduces the dominance of central utilities. This also fosters direct community

DOI: http://dx.doi.org/10.5772/intechopen.1014131

involvement and user-centric energy services that can sustain themselves during power outages. Using on-chain rules, BVPP provides transparent economic incentives that further promote clean energy.

6. Case studies on BVPP

Table 1 provides BVPP use cases; it summarizes the world's famous VPP and blockchain projects along with their pros and cons.

Project	Ref	Focus area	Key technologies	Advantages	Drawbacks
Brooklyn Microgrid (USA)	[9]	BVPP, P2P, energy trading	Private blockchain (LO3 Energy), smart meters, automated contracts	Prosumers earn, local balancing, real-time settlement	Limited scale, regulatory barriers
Power Ledger (Australia/ Global)	[10]	P2P energy trading, VPP	PoA blockchain, Google Cloud, smart contracts, tokenized energy units	Real-time data, automated billing, scalability, renewable integration	Token economics volatility, grid code adaptation
TenneT & Sonnen (Germany/ Netherlands)	[11]	VPP grid support, flexibility	Permissioned blockchain, decentralized storage, cloud analytics	Grid balancing using distributed home batteries prevents curtailment	Complex interoperability, deployment cost
WePower (Estonia)	[12]	Renewables tokenization, energy trading	Ethereum smart contracts, tokenization	Fractional ownership, market access for small producers, transparent RECs	Legal clarity, data standard challenges
Negawatt DR Pilot (USA/ Thailand)	[13]	Demand response, negawatt trading	Hyperledger Fabric, smart contracts	Enables DR as a commodity, creates new markets, opens participation	Needs smart metering, behavioral adoption hurdles
Electron (UK)	[14]	Energy flexibility, grid services	Ethereum blockchain, flexibility trading	Tracks and settles grid flexibility; automates DER aggregation	Participant onboarding, regulatory adaptation
Equigy (Europe)	[15]	Crowd balancing, V2G	Multi-TSO permissioned blockchain, DLT	Aggregation of small-scale EV/battery storage for grid balancing	Complexity in TSO-DSO coordination, interoperability pipeline
Piclo Flex (UK)	[16]	Flexibility trading, ancillary services	API-connected blockchain registry, automated bids	Transparent flexibility procurement, supports DER/DR	Integration with legacy grid infrastructure
Delhi P2P Pilot (India)	[17]	P2P solar trading, local balancing	Blockchain registry, smart meters, mobile interface	Grid stress reduction, real-time trade, instant settlements, prosumer inclusion	Pilot focused, regulatory clarity pending
AutoGrid VPP (India/ Australia)	[18]	VPP, demand response, flexibility	DERM platforms, AI, API/blockchain linkage	Enables millisecond response in VPP, direct market control	Vendor lock-in risk, interoperability limits

Table 1.
Blockchain-enabled VPP and energy trading case studies (2015–2025).

7. Technical challenges and future directions

While blockchain enhances data integrity, privacy-preserving mechanisms, and identity management, it may suffer from Sybil attacks or some times DDoS attacks. A Sybil attack occurs when a malicious actor creates multiple fake prosumer or smart-meter identities in a BVPP to manipulate energy trading or distort demand-response [19]. This undermines consensus, falsifies nega-watt or kilo-watt trades, and threatens grid reliability and market fairness. A DDoS attack, on the contrary, floods the BVPP network, smart meters, or VPP controller with excessive requests, overwhelming communication channels. This ultimately delays block validations, disrupts or may even halt NW/KW trading, and blocks prosumer participation. It increases latency, prevents real-time grid response, and risks destabilizing microgrid operations during peak demand. Additionally, complex cryptographic algorithms may increase system complexity and computational overhead. Moreover, as the number of prosumers and transactions grows, achieving high throughput, low latency, and network efficiency becomes increasingly difficult. End-user acceptance of BVPP also faces many barriers, such as limited blockchain awareness and infrastructure costs [20]. Consumers and utilities hesitate due to unfamiliarity with decentralized models, digital tokens, and data transparency. Furthermore, the absence of standardized regulations and unclear energy trading policies hinders BVPP mass adoption. Regulatory ambiguity affects data ownership, cross-border power transactions, tokenization, and smart contract legality. The major setback in this regard is that implementing BVPP requires interdisciplinary expertise in blockchain, power systems, cyber-security, and data science. The current workforce gap limits efficient deployment, maintenance, and innovation in such frameworks. **Figure 3** shows a clear idea of how BVPP can revolutionize the energy sector.

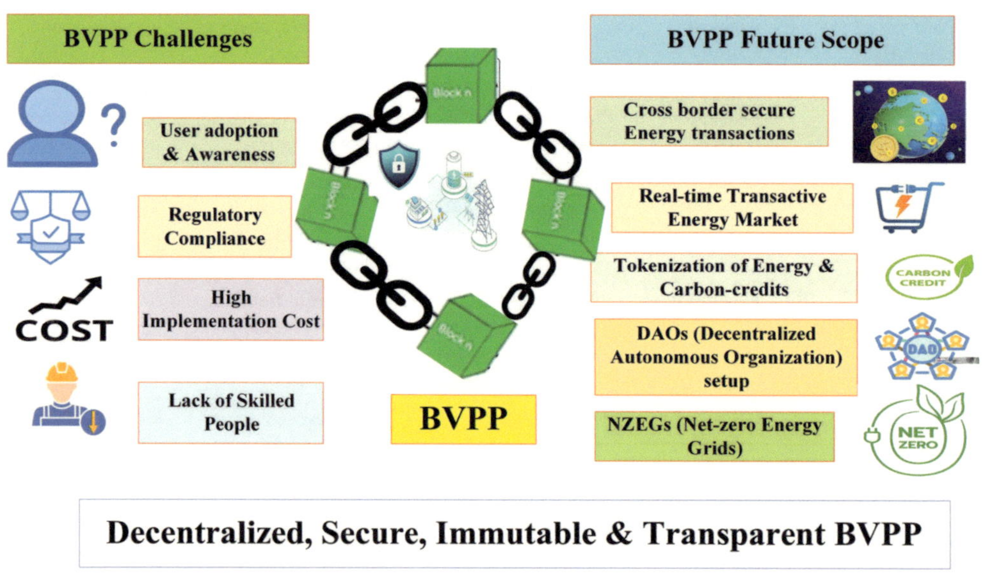

Figure 3.
Blockchain-based virtual power plant (BVPP) for dynamic real-time energy market.

For implementing BVPP for mass adoption, it is crucial to focus on consensus, as the consensus mechanism or chain should be chosen very carefully so that it can handle a huge amount of transactions or provide higher throughput, high scalability, low transaction cost, and energy efficiency [21]. Algorand blockchain, invented by Silvio Micali, uses PPoS and could be a highly suitable fit for BVPP due to its low carbon footprint, high scalability (almost 13,000 TPS), low computational overhead, and low transaction cost (0.001 Algo or 0.000182 USD). Additionally, Algorand claims that it has solved the blockchain trilemma of maintaining security, decentralization, and scalability [22]. PoA [23] and DPoS [24] are other highly suitable consensus mechanisms due to their energy efficiency, scalability, low hardware demands, rapid finality, and fault tolerance. Side chains like Polkadot [25], Skale Network, Ronin, and Polygon [26], which are built over PoS, could be another possible solution for further enhancing blockchain transactions and scalability. Moreover, training programs, academic-industry collaborations, and blockchain-focused energy research initiatives are necessary to build an ecosystem of blockchain-energy-skilled professionals, which is essential to scale BVPP operations in the future [27]. Apart from this, there is a strong need for clear regulatory policies and government intervention [28] to scale BVPP to the global stage.

8. Conclusions

This book chapter represents a significant advancement in modern power generation and management through BVPPs as a transformative solution for future power systems. As the global energy landscape continues to shift from centralized generation toward decentralized, intelligent, cyber–physical, and sustainability-driven grid architectures, BVPPs have emerged as a resilient framework for securely orchestrating and coordinating DERs. These DERs, such as solar PV, wind turbines, battery storage units, EVs, smartly controlled flexible loads, and demand–response technologies, exhibit heterogeneous behaviors and operate under varying levels of controllability. BVPPs leverage advanced real-time optimization, secure communication, and market-driven mechanisms to coordinate these distributed assets in real time, ensuring optimal power balancing, local flexibility procurement, and scalable grid support.

The integration of blockchain fundamentally enhances the operational reliability of VPPs by introducing a decentralized trust layer capable of guaranteeing immutability, transparency, and tamper-proof settlement of energy transactions. Instead of relying on a centralized operator or utility to validate metering data, dispatch signals, and incentive computation, blockchain distributes these responsibilities across a consensus-secured network. This provides strong cryptographic assurance that metering and trading records cannot be forged or manipulated, even under adversarial conditions, an essential requirement for large-scale DER participation. Moreover, BVPP architectures unlock several advanced paradigms for consumer-centric and prosumer-driven ecosystems. Such ecosystems promote democratized energy systems and motivate end users to enhance localized energy management by participating in NW trading: prosumers are incentivized to reduce or shift demand during peak periods. This creates a quantifiable and secure mechanism for demand–response participation, improving grid resilience without requiring additional generation. Moreover, prosumers with surplus energy can transact directly with nearby peers through BVPP. The distributed ledger ensures fair pricing, enhances local energy autonomy, and also alleviates stress on distribution networks by encouraging energy circulation within micro-clusters. Furthermore, BVPP can also promote KW, meaning when DERs generate surplus green energy, BVPP allows seamless export to the utility grid with immediate blockchain-recorded settlement. Thus, in this way, end users can support the grid during peak times, automate ancillary services, increase their revenue collection, and also become active contributors to local balancing strategies. As incentive distribution becomes instantaneous and trust issues are minimized through blockchain, BVPP naturally encourages greater prosumer participation, accelerating renewable adoption and enhancing the flexibility of microgrids and national grids. While challenges persist, such as regulatory alignment, interoperability with legacy grid systems, and scalability of blockchain consensus protocols, continuous technological innovations, Layer-2 scaling, and global pilot demonstrations indicate that BVPPs represent a highly feasible and strategically valuable evolution of modern energy systems.

Future progressive research in AI-driven optimization, standardized protocols, and responsive policy will further unlock the potential of BVPPs. BVPPs are vital to realizing resilient, sustainable, and inclusive future power systems where diverse stakeholders like utilities, prosumers, and communities actively participate in energy generation and management. With the above research, it is clearly visible that BVPPs stand as a cornerstone solution to the challenges and opportunities shaping the future of power generation.

Conflict of Interest

The authors declare no conflict of interest.

Abbreviations

VPP	Virtual power plant
BVPP	Blockchain-based virtual power plant
DERs	Distributed energy resources
NZEGs	Net-zero energy grids
DAOs	Decentralized autonomous organizations
PoW	Proof of work
PoS	Proof of stake
PPoS	Pure proof of stake
ECC	Elliptic curve cryptography
AES	Advance encryption standards
PoA	Proof of authority
dPoS	Delegated proof of stake
PBFT	Practical Byzantine fault tolerance

DOI: http://dx.doi.org/10.5772/intechopen.1014131

Author details

Aakanksha Bedi*†, J. Ramprabhakar† and R. Anand†

Amrita Vishwa Vidhyapeetham, Bengaluru, India

† These authors contributed equally.

*Address all correspondence to: bl.en.r4eee22011@bl.students.amrita.edu

References

[1] Sarathkumar TV, Goswami AK, Khan B, Shoush KA, Ghoneim SSM, Ghaly RNR. Forecasting of virtual power plant generating and energy arbitrage economics in the electricity market using machine learning approach. Scientific Reports. 2025;**15**(1):3812. DOI: 10.1038/s41598-025-03812-6

[2] Zare A, Shafiyi MA. Virtual power plant models and market participation: A deep dive into optimization and real-world applications. Results in Engineering. 2025;**26**:105548. DOI: 10.1016/j.rineng.2025.105548

[3] Yang Q, Wang H, Wang T, Zhang S, Wu X, Wang H. Blockchain-based decentralized energy management platform for residential distributed energy resources in a virtual power plant. Applied Energy. 2021;**294**:117026. DOI: 10.1016/j.apenergy.2021.117026

[4] Andoni M, Robu V, Flynn D, Abram S, Geach D, Jenkins D, McCallum P, Peacock A. Blockchain technology in the energy sector: A systematic review of challenges and opportunities. Renewable and Sustainable Energy Reviews. 2019;**100**:143–174

[5] Li B, Yang F, Qi B, Bai X, Sun Y, Chen S. Research on key technologies of P2P transaction in virtual power plant based on blockchain. IET Smart Grid. 2022;**5**(4):223–233

[6] Gan R, Li R. Contract tokenization in the renewable energy market. SSRN Electronic Journal. 2024;4932043. DOI: 10.2139/ssrn.4932043

[7] Karisma K, Goundar S. Blockchain Reshaping Energy Markets: Regulatory Frameworks for Sustainability-driven Energy Tokens. IntechOpen; 2025. DOI: 10.5772/intechopen1013179

[8] Zakaret C, Peladarinos N, Cheimaras V, Tserepas E, Papageorgas P, Aillerie M, Piromalis D, Agavanakis K. Blockchain and secure element: A hybrid approach for secure energy smart meter gateways. Sensors. 2022;**22**(24):9664. DOI: 10.3390/s22249664

[9] Brooklyn Energy. Available from: https://www.brooklyn.energy [Accessed: 2025-November-01]

[10] Australian Trade & Investment Commission. Available from: https://www.austrade.gov.au/ [Accessed: 2025-November-02]

[11] Sonnen to build Europe's largest virtual home battery storage solution. Available from: https://sonnengroup.com/press/europes-largest-vpp/ [Accessed: 2025-November-02]

[12] WePower revolutionizing green energy exchange. Available from: https://e-estonia.com/weekly-press-review-wepower-on-the-verge-of-revolutionizing-green-energy-exchange/ [Accessed: 2025-November-02]

[13] Thailand clean energy direct purchase. Available from: https://www.trade.gov/market-intelligence/thailand-clean-energy-direct-purchase [Accessed: 2025-November-03]

[14] Orchestrate the full value of DERs using flexibility markets. Available

from: https://electron.net/ [Accessed: 2025-November-04]

[15] Crowd balancing. Available from: https://equigy.com/ [Accessed: 2025-November-04]

[16] The independent marketplace for energy flexibility. Available from: https://www.piclo.com/ [Accessed: 2025-November-05]

[17] Tata Power-DDL peer-to-peer solar energy trading. Available from: https://www.tataworld.com/news/openinside/tata-power-ddl-rolls-out-live-peer-to-peer-solar-energy-trading [Accessed: 2025-November-05]

[18] Virtual power plants: Purpose-built for Australia. Available from: https://www.integratetozero.org/vpp-intelligence-hub/virtual-power-plants-purpose-built-for-australia [Accessed: 2025-November-05]

[19] Douceur JR The sybil attack. In: Proceedings of the International Workshop on Peer-to-Peer Systems. Berlin: Springer; 2002. p. 251–260

[20] Fan M, Tang X, Pei W, Kong L Peer-peer electricity transaction and integrated regulation of VPP based on blockchain. In: IEEE 4th Conference on Energy Internet and Energy System Integration (EI2); 2020 October. IEEE. p. 727–732

[21] Roozbehani MM, Heydarian-Forushani E, Hasanzadeh S, Elghali SB. Virtual power plant operational strategies: Models, markets, optimization, challenges, and opportunities. Sustainability. 2022;**14**(19):12486

[22] Chen J, Micali S. Algorand: A secure and efficient distributed ledger. Theoretical Computer Science. 2019;**777**:155–183

[23] Zeidabadi FA, Dehghani M. POA: Puzzle optimization algorithm. International Journal of Intelligent Engineering and Systems. 2022;**15**(1). DOI: 10.22266/ijies2022.0228.25 [Accessed: 2025-November-16]

[24] Saad SMS, Radzi RZRM. Comparative review of the blockchain consensus algorithm between proof of stake (PoS) and delegated proof of stake (DPoS). International Journal of Innovative Computing. 2020;**10**(2). DOI: 10.11113/ijic.v10n2.272

[25] Burdges J, Cevallos A, Czaban P, Habermeier R, Hosseini S, Lama F, Wood G Overview of Polkadot and its design considerations. arXiv preprint arXiv:2005.13456; 2020

[26] Rana SK, Rana AK, Rana SK, Sharma V, Lilhore UK, Khalaf OI, Galletta A. Decentralized model to protect digital evidence via smart contracts using layer-2 Polygon blockchain. IEEE Access. 2023;**11**:83289–83300

[27] Zhong Y, Ji L, Wang C Discussion of domestic VPP development trend and comparison of domestic and foreign VPP development models and characteristics. In: 2024 IEEE 7th International Electrical and Energy Conference (CIEEC). IEEE; 2024. p. 604–609

[28] NumberAnalytics. Blockchain's impact on energy policy and regulation. Available from: https://www.numberanalytics.com/blog/blockchain-impact-energy-policy-regulation [Accessed: 2025-November-20]

Chapter 3

Perspective Chapter: Robust State Estimation for Future Power Systems – Overcoming the Drawbacks of Traditional State Estimators

Bei Gou, Qing Zhao and David Bogen

Abstract

This chapter presents a comprehensive examination of the limitations of traditional power system state estimation methods, particularly Weighted Least Squares (WLS) and Least Absolute Value (LAV), in the face of growing system complexity, measurement inaccuracies, and stressed operating conditions. Traditional state estimators, originally developed under assumptions of linearity, Gaussian noise, and steady-state operation, increasingly find it difficult to deliver accurate and robust results in modern grids characterized by higher and higher renewable integration, and high-frequency disturbances, including extreme weather events. Through illustrative examples, including a small system case study, this chapter highlights critical issues such as convergence failures, susceptibility to leverage points, and the pervasive problem of negative loads. A new class of estimators is briefly introduced, emphasizing selective measurement use, sparse recovery, and robustness to bad data. Simulation results on large-scale and real-world utility systems demonstrate the superior accuracy, convergence reliability, and bad data rejection capabilities of this new class state estimator. The chapter concludes by identifying key directions for future research in distributed, dynamic, and AI-augmented state estimation, underscoring the need for resilient algorithms in next-generation energy management systems.

Keywords: power system state estimation, weighted least squares, least absolute value, static state estimator plus, robustness, leverage point, objective function

1. Introduction

State estimation is an analytical application designed to determine the probable operating state of a power system transmission network [1]. This methodology, originally introduced by Schweppe in the 1960s [2], has since become a cornerstone in the monitoring and analysis of power systems. Traditional State Estimation (SE)

IntechOpen

algorithms reliably estimate the transmission system's operating state when it operates under steady-state conditions with no significant frequency deviations [3], which was historically a reasonable assumption.

However, traditional SE algorithms are not without limitations. The precision of the State Estimator results heavily depends on the single-phase model and the availability and precision of system measurements, such as voltage magnitude, power flow measurements on branches, and real and reactive power injections at each bus [1–3]. Over time, the transmission system has expanded significantly, becoming approximately 5–6 times larger, and it no longer consistently operates in a steady-state condition due to the penetration of renewable energy and implementation of the power market [3]. To illustrate this point, consider load growth between 2020 and 2025: the average increase in the U.S. was around 3%, whereas Texas is projected to see a 30% increase. This rapid evolution in grid size and load dynamics has diminished the reliability and accuracy of traditional state estimator approaches.

Coinciding with this rapid growth, the power system is undergoing substantial modernization, with new energy generation methods being deployed. These include solar and wind power, often integrated with energy storage solutions like batteries. Since these devices rely on power electronics, they respond to changes in the power system extremely quickly. Concurrently, the availability of measurements for the power system has increased; today's systems have a greater percentage of real-time measurements for the state estimator compared to the 1960s, which relied more on real-time and pseudo-measurements [1, 4, 5].

1.1 Importance of state estimation in power systems

Today, State Estimator is an essential application for real-time operations of the power grid. Its critical importance is highlighted by the following significant past events:

- The 2011 Southwest United States Blackout, for instance, originated from a single transmission line failure and rapidly escalated into a cascading grid collapse, resulting in an estimated $100 million economic impact. This event powerfully demonstrates the profound societal consequences of grid failures [6].

- More recently, the near collapse of the ERCOT Grid in Texas during Winter Storm Uri underscored the inherent fragility of power systems [6]. Reports indicate the ERCOT system was merely 4 minutes and 37 seconds away from total system failure [7], leading to restoration times measured in days rather than hours or minutes.

These types of events emphasize the necessity of monitoring, analyzing, and reporting power system conditions within minutes, enabling grid operators to take swift action to prevent catastrophic failures.

Beyond grid reliability, the accuracy of state estimation also carries significant financial implications. Many regions determine energy transaction prices based on locational pricing, which in turn is derived from planned and existing real-time operating state of the transmission grid [8]. This necessitates that the state estimator consistently provides accurate and reliable results. What was once primarily viewed as a reliability tool has now also become a crucial financial instrument.

1.2 Problems suffered by traditional state estimators

Despite significant advancements in measurement and communication technologies, the core algorithmic foundation of state estimation has remained largely unchanged in the power industry over the past 50 years. Traditional methods—especially Weighted Least Squares (WLS) and Least Absolute Value (LAV)—were built on assumptions that are increasingly incompatible with modern power system realities. These include the use of a single-phase model, P/Q/V measurement sets, non-simultaneous measurements, and steady-state operation at a single frequency [2, 3]. Such assumptions imply balanced three-phase conditions and perfect network models based on positive sequence notation, which rarely hold true in today's power grids [3].

Traditional estimators also assume Gaussian-distributed, independent measurement noise and perfect knowledge of network topology and parameters. However, in practice, measurement errors are often non-Gaussian, and system models frequently contain inaccuracies. Factors such as high penetration of renewables, dynamic operating conditions, cyber threats, and hardware failures introduce complexities that traditional estimators were not designed to handle [6].

Furthermore, traditional estimators lack inherent mechanisms to distinguish between reliable and erroneous data. WLS attempts to fit all available measurements, including bad ones, which leads to biased estimates. LAV, though more robust to outliers, still fails when bad data occur at leverage points—measurements with high influence due to their location or assigned weight. These estimators are particularly vulnerable to corrupted data, which can dominate the solution and go undetected.

Another core limitation lies in their reliance on linearization. Traditional state estimators use iterative Gauss-Newton-type methods that require well-conditioned Jacobian matrices and good initial guesses to converge. In stressed or ill-conditioned systems—such as during contingencies or with topology errors—these methods may diverge or converge to physically implausible solutions. This vulnerability was notably observed in the 2003 blackout, where a topology error caused the state estimator to crash, and operators had to manually intervene due to the system's failure to provide a valid solution.

Bad data detection in traditional frameworks is generally performed after the initial estimate using residual-based methods, such as the largest normalized residual test. However, these techniques depend on the availability of a converged solution and are ineffective in the presence of severe data corruption or leverage points. In many cases, especially under significant errors, the estimator cannot even produce an initial solution, thus leaving no foundation for subsequent bad data identification [6, 9].

While it is often assumed that increasing the number of measurements improves estimator performance, this can introduce more bad data or communication-related anomalies. Modern measurement infrastructures use internet-based protocols, where failures in a shared communication path can simultaneously affect multiple RTUs. As a result, large portions of the power system can experience missing or delayed measurements. Additionally, measurements from devices such as Intelligent Electronic Devices (IEDs) and Phasor Measurement Units (PMUs) are received at varying rates and time scales, leading to synchronization and skew challenges that traditional estimators are ill-equipped to address.

Even when data is available, traditional estimators can produce physically unrealistic results. A notable example is the generation of negative loads, where load buses appear to be generating power—an artifact of unconstrained optimization without physical validation. These anomalies complicate downstream applications like contingency analysis, voltage stability assessment, and pricing.

In summary, the foundational weaknesses of traditional state estimators include:

- Obsolete modeling assumptions [2, 3],
- Sensitivity to bad data and leverage points [6, 9],
- Dependence on convergence and initialization [10],
- Ineffectiveness in detecting hidden or structured data errors, and
- Inability to enforce physically realistic constraints.

These challenges highlight the urgent need for structurally robust and adaptive estimation frameworks capable of operating reliably in the face of complex, uncertain, and dynamic system conditions.

Traditional state estimators have been widely adopted and successfully implemented by many power utilities. Some companies, even without regulatory obligation, have expressed a strong interest in running their state estimators in real-time environments. However, despite these efforts, limitations in the accuracy, robustness, or convergence of traditional estimators have prevented some utilities from fully integrating them into their real-time operations. This highlights the performance gap that still exists between current state estimation tools and the stringent demands of real-time grid monitoring and control.

1.3 Traditional state estimators for power transmission grids

Among traditional state estimators, weighted least square state estimator and the least absolute value state estimator are the most widely applied in power systems, which are also the two state estimators received the most extensive research and investigation since the first date when the state estimator was introduced to power systems in the 1960s.

The ability to analyze and study the behavior of the transmission system in real-time has become a critical requirement as power system complexity escalates. As problems with traditional state estimation methods persist and as the transmission system becomes more and more complex with instantaneously changing operating states, new algorithms for state estimation are imperative.

Traditional Power System State Estimation (SE) is the core application providing real-time “snapshots” of complex bus voltages, crucial for grid operators. It enables critical functions like contingency analysis and optimal power flow. The process models the network *via*

$$z = h(x) + e$$

where x is the state vector (voltages/angles), z represents SCADA measurements (P, Q, V, flows), and e is assumed Gaussian noise [2, 4, 5, 11].

Below is a summary of main types of traditional state estimators.

1.3.1 Weighted Least Squares (WLS)

The Weighted Least Squares (WLS) state estimator is the most widely used algorithm in power system control centers. It estimates the system’s bus voltages by minimizing the sum of weighted squared differences between actual measurements (e.g., power flows, injections, and voltages) and their corresponding model-based estimates. WLS provides a statistically optimal estimate under the assumption of Gaussian measurement noise. WLS estimator minimizes the weighted least squares cost

$$J = \|z, -, h, (x)\|_R^2 = (z - h(x))^T R^{-1} (z - h(x))$$

using Gauss-Newton iterations. This ensures quadratic convergence with good initial guesses. Efficiency is enhanced by techniques like fast-decoupled WLS and equality-constrained WLS for embedding hard constraints. Post-solve, bad-data processing (e.g., χ^2, largest nominalized residuals (LNR) tests) flags and isolates faulty measurements for re-estimation [2, 11].

Historically, state estimators assumed steady-state operation, balanced single-phase modeling, Gaussian noise, and sequential SCADA scans, making it practical for 1960s computing. However, rising PMU penetration, renewable variability, and cyber threats now necessitate more robust, dynamic, and distributed extensions to this foundational algorithm [1, 9].

1.3.2 Least absolute value (LAV)

The Least Absolute Value (LAV) state estimator was introduced as a more robust alternative to WLS, particularly in scenarios where measurement errors deviate from the Gaussian assumption or when outliers and bad data are present. Unlike WLS, which minimizes the sum of squared residuals, LAV minimizes the sum of absolute residuals:

$$J = \|z, -, h, (x)\|_1$$

This formulation reduces the influence of large residuals, thereby improving resilience to bad or corrupted data. However, LAV leads to a non-differentiable and non-quadratic optimization problem, which is more challenging to solve than WLS. Early implementations used linear programming or iteratively reweighted least squares, but convergence could be slow or unreliable for large-scale systems. Despite these difficulties, LAV has been shown to outperform WLS in the presence of gross measurement errors, and has motivated continued research into robust and efficient solution techniques for practical application in power system state estimation [11].

The main obstacles preventing the widespread application of the LAV state estimator in power systems are its slow computational speed and its vulnerability to leverage points. The LAV estimator tends to produce near-zero residuals for bad leverage points, making them extremely difficult to detect. Despite significant research efforts, the leverage point problem remains an open challenge.

1.4 Motivation for advanced solutions

As can be seen from the discussion so far, as the Power System continues to grow, the need for better and faster analysis applications will continue to be required. And this means the bridge between the 'real world' and the analytical world will continue to be required. This bridge will continue to be the state estimator. To keep up with grid modernization challenges and rapid changes in the operating states of the Power System, new solutions will be required that can analyze the current and predict the future state of the Power System with increasing levels of accuracy and dependability. As the financial impact of failure of these applications increases, the importance of operating the Power System becomes critical and failure or incorrect results will not be tolerated. The future of these applications will depend on newer and better methods; some are currently in their infancy, and others are yet to be developed.

1.4.1 Brief introduction to static state estimator plus (SSE+)

Gou and Shue [9] proposed a new state estimator, Static State Estimator (SSE+), designed to address several fundamental limitations inherent in traditional power system state estimators. The key advantages of SSE+ are summarized as follows:

1. Nonlinear equation handling: Unlike traditional estimators that rely on linearizing nonlinear measurement equations (e.g., *via* Gauss-Newton iterations), SSE+ solves the original nonlinear formulation directly. This approach avoids approximation errors introduced by linearization and results in a more accurate and stable estimation process.

2. Robustness to bad data: Traditional WLS estimators are sensitive to bad measurements, which can arise from gross measurement errors, communication failures, or incorrect topology or parameter information. SSE+ is designed with built-in bad data rejection capabilities, effectively mitigating the impact of such errors on the final state estimate.

3. Improved convergence: SSE+ avoids many of the convergence issues that affect both WLS and LAV estimators, particularly in stressed or ill-conditioned systems. By employing a new objective function formulation and leveraging compressive sensing principles, SSE+ can converge to meaningful solutions even when traditional estimators fail.

4. Basis set and sparse residual utilization: SSE+ introduces a novel use of sparse linear residual equations derived from the measurement model to identify reliable measurement subsets, termed the basis set. This allows for targeted estimation and outlier exclusion without relying on iterative residual tests or reweighting procedures.

Overall, SSE+ represents a significant advancement in state estimation, offering improved accuracy, robustness, and convergence across a wide range of operating conditions.

2. Drawbacks of traditional state estimators

Despite their longstanding role in Energy Management Systems (EMS) operations, traditional state estimators like WLS and LAV suffer from several limitations that undermine their reliability, particularly in adverse or error-prone environments.

2.1 Fundamental assumptions of WLS state estimator

The Weighted Least Squares algorithm is a traditional optimization-based technique for power system state estimation. It operates by minimizing the sum of weighted squared measurement residuals, defined as the discrepancy between measured values and those predicted by the state variables:

$$J(x) = \sum_{i=1}^{m} w_i (z_i - h_i(x))^2$$

where $z_i$is the measured value, $h_i(x)$ is the nonlinear function of the state variables x, and w_i is the weight assigned based on measurement accuracy.

This formulation is grounded in statistical estimation theory and assumes that measurement errors are zero-mean, independent, identically distributed, and follow a Gaussian (normal) distribution. Under these assumptions, the WLS estimator yields the maximum likelihood estimate of the system state.

The success and accuracy of WLS rely on several critical assumptions [12]:

1. Measurement noise is Gaussian: The algorithm is statistically optimal only if measurement noise follows a normal distribution. If errors are heavy-tailed or include outliers, WLS becomes sensitive and can produce biased or erroneous state estimates.

2. Independence and homoscedasticity: WLS assumes that measurement errors are uncorrelated and that their variances are known and fixed. Correlated measurement errors or inaccurate weight assignment can significantly degrade estimator performance.

3. Accurate system model: The functions $h(x)$ depend on the system model, including network topology and branch parameters (e.g., line impedances, transformer tap settings). WLS assumes this model is perfectly known. In practice, topology or parameter errors (e.g., switch status mistakes or outdated line parameters) can introduce modeling errors, leading to systematic estimation bias or even divergence.

4. Good initial guess: For nonlinear WLS formulations, a good initial estimate is often required for iterative solvers such as Gauss-Newton or Newton-Raphson to converge.

While WLS has been widely adopted in Energy Management Systems (EMS) due to its mathematical simplicity and efficient computation, its performance can deteriorate in real-world scenarios where the above assumptions are no longer valid, which are normally caused by different types of bad data, gross measurement errors, and system model inaccuracies. These limitations have motivated the development of more robust state estimation techniques that can tolerate such deviations from ideal assumptions.

2.2 Vulnerability to bad data

Traditional WLS is especially vulnerable to bad data because it implicitly assumes that all measurements are reliable and minimizes the sum of the weighted squared residuals. When bad data are present, the estimator compromises by adjusting the state vector to fit both accurate and erroneous measurements. This often results in errors being spread across the entire estimated state rather than being isolated. Consequently, WLS estimates can suffer from:

- Biased voltage magnitudes and angles,
- Inaccurate power flow calculations,
- Misidentification of system events or contingencies, and
- Misleading situational awareness for system operators.

Even a small number of undetected bad measurements can significantly distort the estimate—especially in stressed or weakly observable systems. While residual-based techniques, such as the largest normalized residual test, are often applied post-estimation to detect bad data, they come with notable limitations:

- They can fail to detect errors located at leverage points—measurements that exert disproportionate influence on the estimate (further discussed in Section 2.4).
- They rely on statistical thresholds and may be ineffective when errors are correlated or when the noise deviates from Gaussian assumptions.
- They operate as post hoc tools, providing no influence on the estimation process unless bad data are actively removed and the estimator is rerun.

In response to WLS's shortcomings, the Least Absolute Value (LAV) estimator was introduced as a robust alternative. While LAV is more resistant to large residuals due to its use of the L1-norm cost function, it has a critical limitation: its breakdown point is zero when bad data appear at leverage points [13]. As shown by Rousseeuw and others, LAV fails to reject even a single bad leverage point, which can dominate the solution and render the estimate unreliable. Therefore, despite its improved handling of gross errors compared to WLS, LAV remains vulnerable to strategically placed or high-impact bad measurements.

Furthermore, iterative bad data removal, such as excluding suspicious measurements followed by re-estimation, can become computationally expensive and still miss coordinated or hidden attacks. Because the power system is a nonlinear system, WLS and LAV state estimators may consider an error-free measurement as bad data, which mistakenly has the largest normalized residual. These challenges underscore the need for more robust estimation frameworks that not only reduce sensitivity to bad data but also incorporate mechanisms for rejecting unreliable measurements during the estimation process itself.

2.3 Convergence challenges under stressed conditions

Under stressed operating conditions, such as periods of high system loading, voltage instability, contingencies, or following a disturbance, the convergence behavior of the traditional WLS state estimator often deteriorates. One of the primary technical reasons is that the Jacobian matrix of the nonlinear measurement equations, which is central to the Newton-based iterative solution, becomes ill-conditioned or nearly singular. This occurs when small changes in the state variables produce very small or poorly scaled changes in the measurement residuals, indicating a loss of numerical sensitivity and observability.

Since WLS typically relies on Newton's method or the Gauss-Newton approximation to solve the nonlinear optimization problem, convergence heavily depends on the conditioning of the Jacobian and the proximity of the initial guess to the true solution. When the system is stressed or presents with topology errors [14, 15]:

- The Jacobian matrix loses rank, causing the linear system at each iteration to become unstable or unreliable.
- The solution trajectory can oscillate between points without making meaningful progress.
- The algorithm may take tiny step sizes, resulting in premature termination before reaching a true minimum.
- In severe cases, the iteration process may diverge entirely, failing to produce any valid estimate.

These convergence failures are especially problematic because they tend to occur precisely when accurate state information is most critical—during emergencies or system anomalies. A 2020 report [16] from the MIT Energy Initiative observed that traditional state estimators "have trouble estimating a system state during unusual or emergency conditions—unfortunately, when they are most needed."

The convergence issue is not merely a theoretical inconvenience; it poses practical risks. An estimator that fails to converge can leave operators blind to critical system changes, delaying emergency responses and corrective actions. In some utilities, this has led to operators disabling bad data detection routines or skipping iterations to force convergence, further compromising the reliability of situational awareness.

These limitations emphasize the need for robust, non-divergent state estimation algorithms that can maintain stability and accuracy even under stressed or degraded operating conditions. Such methods must be capable of handling nonlinearities more gracefully, avoiding dependency on favorable initial conditions, and tolerating model inaccuracies without collapse.

2.4 The issue of leverage points

In statistical estimation and regression theory, leverage points are data points that have a disproportionately large impact on the estimated solution. In the context of power system state estimation, leverage points correspond to measurements whose influence on the solution is significantly higher than that of others, due to either network structure, measurement type, or weight assignment. These points can unduly shape the outcome of the state estimation process, especially when they are erroneous, making them particularly dangerous if undetected.

Common examples of leverage points in power systems include [17]:

- Injection measurements at highly connected buses: Buses with a large number of incident branches have a wide influence on the network. Errors in injection measurements at these buses can distort the voltage estimates across large portions of the system.

- Measurements on very low-impedance lines: Because low-impedance lines exhibit minimal voltage drop, any flow or current measurement across such lines tightly constrains the voltage difference between connected buses. As a result, an error in such a measurement can propagate strongly through the Jacobian matrix, distorting the entire state vector.

- Measurements assigned excessive weights: Perhaps the most notable example is zero-injection buses, which are often treated as exact measurements. These indicate that a bus has neither generation nor load, leading operators to assign extremely high weights (or effectively infinite precision) to enforce the zero-injection constraint.

While this might seem reasonable under ideal modeling assumptions, zero-injection buses are highly susceptible to becoming hidden sources of error. If any branch connected to a zero-injection bus contains a topology error (e.g., incorrect breaker status) or parameter error (e.g., wrong line impedance), then the actual power injection will no longer be zero. However, because the zero-injection constraint is treated as exact, the estimator forcefully tries to satisfy it, adjusting nearby voltage estimates in a biased manner.

This creates a hidden bad data condition—the measurement is incorrect, but residual-based detection fails because the weight is so high that the residual is suppressed. Standard bad data detection methods, such as the largest normalized residual tests, become ineffective at identifying errors at leverage points, especially when the measurement dominates the estimation process.

Moreover, the presence of such high-leverage bad data can mask other errors, compromise convergence, and result in a false sense of accuracy in the estimated state. Leverage points can be especially problematic in sparse or weakly observable networks, where each measurement carries more influence.

To mitigate these risks, alternative approaches are needed, such as:

- Robust estimators that reduce sensitivity to outliers and high-leverage data,

- Influence analysis to identify and monitor measurements with excessive impact on the solution,

- Adaptive weighting that accounts for the possibility of modeling errors, even in traditionally trusted measurements like zero-injections.

In summary, leverage points represent a critical blind spot in traditional state estimators whose objective functions are based on the residual magnitudes, including WLS and LAV-based estimation, one that can severely undermine both the accuracy and reliability of state estimates—particularly in the presence of hidden model errors.

2.5 Inability to accurately recover true state values

A fundamental limitation of the Weighted Least Squares estimator is its inherent tendency to fit all measurements simultaneously, including those that are flawed. Because WLS minimizes the total weighted sum of squared residuals, it does not inherently distinguish between correct and incorrect data unless additional mechanisms, such as bad data detection and removal, are explicitly implemented. As a result, the estimated state variables often reflect a compromise between accurate and inaccurate information.

This compromise leads to several undesirable consequences [9]:

- True system states become biased: When even a few bad measurements are present, they can pull the solution away from the true values. The estimated voltage magnitudes and angles may no longer represent the actual operating condition of the system.
- Bad data are absorbed rather than rejected: Since WLS gives all measurements a role in shaping the estimate, erroneous values are not excluded by default. Instead, their errors are redistributed across the network, contaminating the state estimate in subtle but significant ways.
- Recovered measurement values are distorted: The measurement values recovered by projecting the estimated state back through the measurement functions $h(x)$ may no longer resemble the true system behavior. This includes power injections, flows, and bus voltages, which can deviate from actual values even when the residuals appear small.

For example, consider a case where a power injection at a bus is erroneously recorded as 500 MW due to a sensor malfunction, while the actual value is 200 MW. WLS will attempt to adjust the surrounding voltage angles and magnitudes to minimize the mismatch, rather than reject the measurement. This typically results in:

- A distorted voltage profile around the erroneous bus,
- Artificial shifts in power flows in adjacent lines,
- And propagation of estimation error to other buses—even those with no measurement error themselves.

Critically, the flawed measurement is never truly corrected. Instead, the estimator masks the problem by diffusing it, potentially lowering the residual to within acceptable limits but corrupting the overall solution. The resulting estimate may still appear to “fit the data” in a statistical sense, but it no longer reflects the physical reality of the system.

This inability to recover the true state is particularly dangerous in scenarios where precise estimates are required, such as:

- Contingency analysis,
- Voltage stability assessment,
- Real-time control decisions, and
- Market settlement calculations.

Moreover, in large-scale systems with thousands of measurements and states, these subtle biases can accumulate and lead operators to misinterpret system conditions, such as underestimating load, overlooking overloads, or misidentifying the location of a disturbance.

To address these limitations, modern estimation methods must move beyond WLS's "fit-all" approach. Instead, they should be capable of identifying and isolating erroneous data, or better yet, inherently rejecting the influence of flawed measurements during the estimation process. Only then can the state estimator reliably recover true system values and provide accurate situational awareness.

2.6 Negative loads in state estimation

In WLS-based state estimation widely used in power systems, it is common to encounter the phenomenon of negative loads, where the estimated power injection at load buses appears positive, suggesting that the bus is generating power rather than consuming it. These physically unrealistic results arise because WLS does not impose explicit constraints on the direction or sign of power injections during the estimation process. Instead, the algorithm simply adjusts the state variables to minimize the residuals, regardless of whether the resulting estimates conform to known operational characteristics.

As a consequence, WLS may produce dozens or even hundreds of negative loads in a moderately sized power system, especially under noisy measurement conditions or in the presence of bad data. These outcomes conflict with the basic understanding of system operation, where load buses are expected to consume power. Furthermore, negative loads can:

- Mislead operators by distorting load distribution across the network,
- Affect contingency and reliability analysis,
- Complicate post-processing and validation procedures.

In contrast, Static State Estimation Plus (SSE+) addresses this issue by significantly reducing the occurrence of negative loads, often limiting them to fewer than 10 in practical implementations. This improvement stems from SSE+'s strategic selection of a clean basis set, composed of measurements that are consistent, well-behaved, and physically trustworthy. By focusing the estimation process on this reliable subset of data, SSE+ avoids overfitting to flawed or noisy measurements that could otherwise distort the results.

Rather than attempting to fit all data equally, SSE+ leverages the structure and redundancy in the measurement set to solve the estimation problem using only the most reliable information. This inherently suppresses anomalous artifacts like negative loads and leads to state estimates that are not only mathematically consistent but also physically meaningful and operationally realistic.

The ability to minimize the occurrence of negative loads underscores the robustness of SSE+ and its superiority over traditional WLS in producing state estimates that truly reflect the actual conditions of the power system. This capability is especially important in real-time applications, where credible and interpretable estimates are essential for safe and reliable system operation.

3. Illustrative examples of WLS and LAV limitations

To demonstrate the performance limitations of traditional estimators, we present a small but representative test case. This example uses a simplified 4-bus DC model with various measurement scenarios. It reveals how bad data corrupt the solution under the WLS method and how such estimators struggle with identification and correction.

3.1 Case study: Bad data impact on state estimation

3.1.1 WLS state estimator as average estimator

We want to use this small example to show that the WLS state estimator is essentially an average estimator.

A scalar variable x is defined, and we have three noisy linear measurements:

$$z_i = x + e_i, i = 1, 2, 3$$

$$Z = \begin{bmatrix} 1.0 \\ 2.0 \\ 1.5 \end{bmatrix}$$

Noise variances: $\sigma_1^2 = 1, \sigma_2^2 = 4$, $\sigma_3^2 = 1$.

WLS state estimator reached an estimate of x:

$$x_{WLS} = 1.3.$$

Let us calculate the weighted average of all the measurements:

$$\hat{x} = \frac{\frac{z_1}{\sigma_1^2} + \frac{z_2}{\sigma_2^2} + \frac{z_3}{\sigma_3^2}}{\frac{1}{\sigma_1^2} + \frac{1}{\sigma_2^2} + \frac{1}{\sigma_2^2}} = \frac{1.0 + 0.5 + 1.5}{1 + 0.25 + 1} = \frac{3.0}{2.25} = 1.3$$

It is obvious that the WLS state estimator is basically an average estimator, which calculates the weighted average of all the measurements, including the bad ones.

The next example shows how the bad measurements affect the solution of the WLS state estimator.

3.1.2 4-bus DC system

For simplicity, we use a 4-bus power system to show the performance of WLS, LAV, and SSE+ with bad data. The system information is given below:

- A 4-bus system with buses labeled 1 through 4.
- All transmission line reactance $x = 1.0$p.u.
- Bus 1 is the reference bus with $\theta_1 = 0.0^o$.
- The state variables to estimate are the voltage phase angles at buses 2, 3, and 4.
- Measurements are given as real power flows along lines or at buses.

The single-phase graph of this system is given as follows (**Figure 1**):

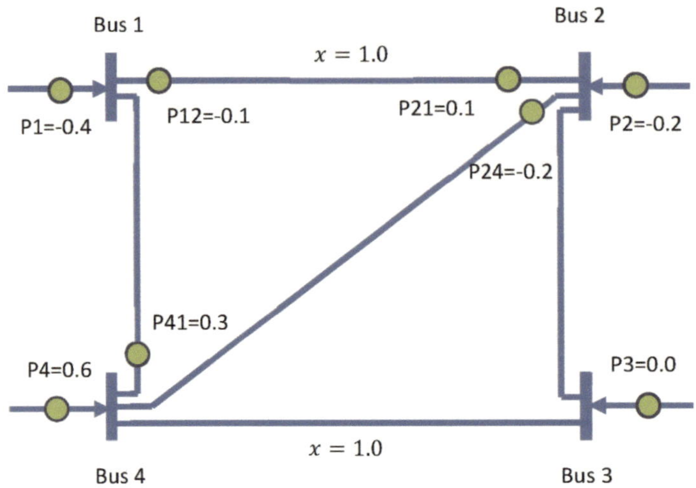

Figure 1.
4-bus DC power system.

For the given set of measurements, we have the Jacobian matrix H as follows:

$$H = \begin{bmatrix} 1 & -1 & 0 & 0 \\ -1 & 1 & 0 & 0 \\ 0 & 1 & 0 & -1 \\ -1 & 0 & 0 & 1 \\ 2 & -1 & 0 & -1 \\ -1 & 3 & -1 & -1 \\ 0 & -1 & 2 & -1 \\ -1 & -1 & -1 & 3 \\ 1 & 0 & 0 & 0 \end{bmatrix}$$

The real power measurements are given in per-unit values:

$$Z = \begin{bmatrix} -0.1 \\ 0.1 \\ -0.2 \\ 0.3 \\ -0.4 \\ -0.2 \\ 0.0 \\ 0.6 \\ 0.0 \end{bmatrix}$$

$$\theta^{true} = \begin{bmatrix} 0.0 \\ 0.1 \\ 0.2 \\ 0.3 \end{bmatrix}$$

3.1.2.1 Case 1: No bad data

When the measurements contain only noise, WLS performs very well, and it can reach the true state values:

$$\theta^{estimated} = \begin{bmatrix} 0.0 \\ 0.1 \\ 0.2 \\ 0.3 \end{bmatrix}$$

$$\theta^{true} = \begin{bmatrix} 0.0 \\ 0.1 \\ 0.2 \\ 0.3 \end{bmatrix}$$

The results show that WLS state estimator is able to achieve accurate state variables when only noise is present in the measurements.

3.1.2.2 Case 2: Single bad measurement

We now inject a gross error into the measurement P_2 by forcing (**Figure 2**) $P_2 = 0.5$:

3.1.2.2.1 WLS state estimator

WLS state estimator reached the following values for the state variables:

$$\theta^{estimated}_{WLS} = \begin{bmatrix} 0.0 \\ 0.2181 \\ 0.1467 \\ 0.2770 \end{bmatrix}$$

Using the obtained estimate of state variables, we calculated the measurement recovered values and the residuals (**Table 1**).

It is obvious to see that the residuals of all measurements are affected because all the state variables are biased due to this single bad measurement.

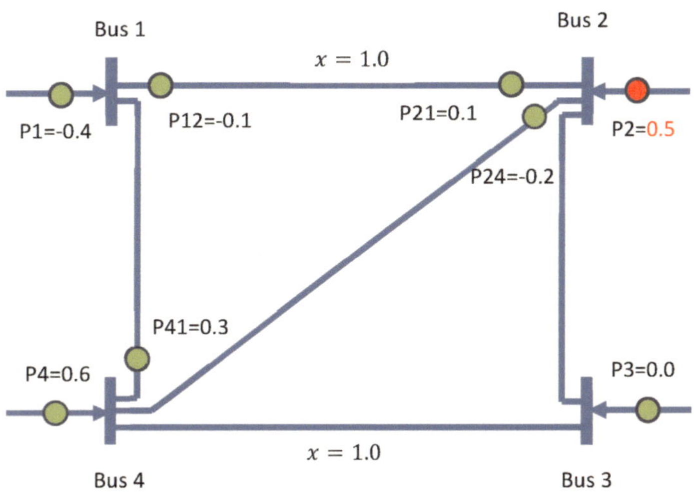

Figure 2.
4-bus system with a single bad measurement.

	P12	P21	P24	P41	P1	P2	P3	P4
Measurements	−0.1	0.1	−0.2	0.3	−0.4	0.5	0	0.6
Recovered measurements	−0.2181	0.2181	−0.0588	0.277	−0.4951	0.2307	−0.2016	0.466
Residuals	0.1181	−0.1181	−0.1412	0.023	0.0951	0.2693	0.2016	0.134

Table 1.
WLS state estimator recovered values and obtained residuals of the measurements.

3.1.2.2.2 SSE+ (static state estimator plus)

SSE+ was introduced in Ref. [9], which optimizes the selected basis set of measurements such that all measurements in the basis set are error-free. The basis set was formed by measurements P_{12}, P_{24}, P_3, highlighted in yellow (**Figure 3**).

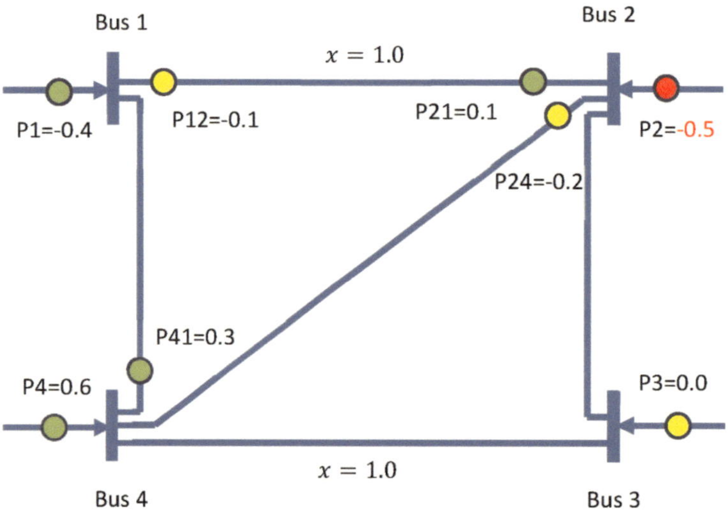

Figure 3.
4-bus system with SEE+ basis set.

SSE+ used the basis set measurements to solve the state variables. It exactly solved the state variables because the basis set is error free.

$$\theta_{SSE+}^{estimated} = \begin{bmatrix} 0.0 \\ 0.1 \\ 0.2 \\ 0.3 \end{bmatrix}$$

Using the obtained estimate of state variables, we calculated the measurement recovered values and the residuals (**Table 2**).

We can see that all measurements except the bad measurement P_2 are exactly recovered because the estimate of the state variables is accurate. Even for the bad measurement P_2, its true value -0.2 is exactly recovered.

3.1.2.3 Case 3: Multiple bad measurements

In this scenario, we investigate the impact of multiple bad measurements on the state estimation results. Specifically, we intentionally corrupt three measurements in the system:

	P12	P21	P24	P41	P1	P2	P3	P4
Measurements	−0.1	0.1	−0.2	0.3	−0.4	0.5	0	0.6
Recovered measurements	−0.1	0.1	−0.2	0.3	−0.4	−0.2	0	0.6
Residuals	0	0	0	0	0	0.7	0	0

Table 2.
SSE+ recovered measurement values and obtained measurement residuals.

- $z_2 = 0.5$
- $z_3 = 0.0$
- $z_4 = -0.3$

The simultaneous presence of multiple bad measurements makes it more difficult for traditional state estimators to isolate and reject errors. WLS, in particular, attempts to fit all data points, including the bad ones, which often leads to a distorted estimate. The influence of these errors may spread across the entire state vector, and residual-based detection becomes less effective (**Figure 4**).

In contrast, SSE+ identifies these measurements as inconsistent by focusing on minimizing the *number* of abnormal residuals rather than minimizing the objective function formed by their *magnitudes*. This allows SSE+ to reject the bad data with the highest possibility and recover a more accurate and physically meaningful estimate.

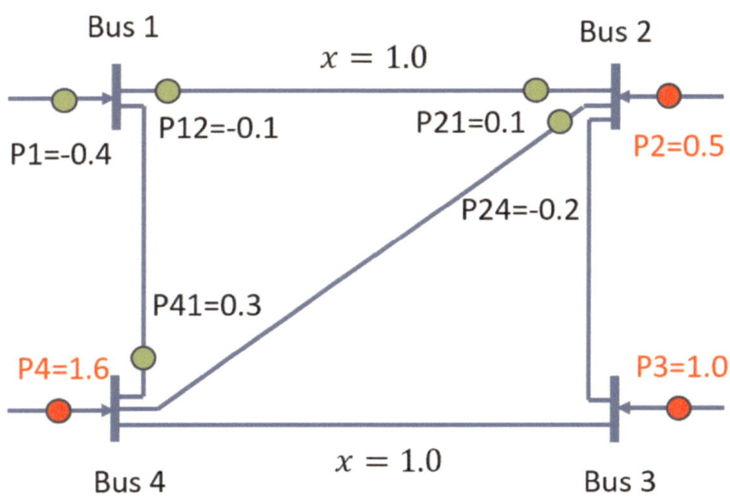

Figure 4.
4-bus power system with three bad measurements.

3.1.2.3.1 SSE+ (static state estimator plus)

Table 3 shows that SSE+ can minimize the influence of bad measurements because only one state variable is biased and two measurements received non-zero residuals. However, if all three bad measurements are rejected, the entire system becomes unsolvable.

Variable			True values				Estimate values
θ_1			0.0				0.0
θ_2			0.1				0.1
θ_3			0.2				0.7
θ_4			0.3				0.3
Measurements	**P12**	**P21**	**P24**	**P41**	**P1**	**P2**	**P3**	**P4**
Measurements values	−0.1	0.1	−0.2	0.3	−0.4	0.5	1	1.6
Recovered measurements	−0.1	0.1	−0.2	0.3	−0.4	−0.7	1	0.1
Residuals	0	0	0	0	0	1.2	0	1.5

Table 3.
SSE+ recovered measurement values and obtained measurement residuals for multiple bad measurements.

3.1.2.3.2 WLS state estimator

In this situation, the solution becomes even more distorted. The voltage estimates are now significantly removed from their true values. The WLS estimator spreads the impact of errors across the solution space to satisfy the optimization goal (**Table 4**).

This demonstrates a critical vulnerability: WLS cannot isolate or recover the true state values once multiple errors are present.

3.2 Analysis of leverage points and biased solutions

If a leverage point measurement is corrupted, WLS and LAV state estimators are forced to "fit" it, regardless of its validity.

This behavior makes WLS and LAV particularly vulnerable in practice because:

- Residuals for leverage points are suppressed,
- Bad data in such locations go undetected,
- Bias accumulates silently in the estimated state.

Variable			True values					Estimate values
θ_1			0.0					0.0
θ_2			0.1					0.2840
θ_3			0.2					0.5438
θ_4			0.3					0.5568
Measurements	**P12**	**P21**	**P24**	**P41**	**P1**	**P2**	**P3**	**P4**
Measurements values	−0.1	0.1	−0.2	0.3	−0.4	0.5	1	1.6
Recovered measurements	−0.284	0.284	−0.2728	0.5568	−0.8407	−0.2488	0.2469	0.8426
Residuals	0.184	−0.184	0.0728	−0.2568	0.4407	0.7488	0.7531	0.7574

Table 4.
WLS state estimator recovered measurement values and obtained measurement residuals for multiple bad measurements.

The following example gives the responses of the LAV state estimator and SSE+ to leverage point measurements.

Consider a simple regression model given below [2]:

$$z_i = A_{i1}x_1 + A_{i2}x_2 + e_i \qquad i = 1, \bullet\bullet\bullet, 5$$

Where A_{ij}, represents the jth entry in vector A_i as defined above. The associated observation data are given in the table below. $SD = 0.001$.

3.2.1 Test 1: Add a bad data on a normal measurement

The measurement configuration (a bad data is added on the third measurement) (**Table 5**).

i	z_i	A_{i1}	A_{i2}
1	14.9954	1.0	1.0
2	18.0053	2.0	1.0
3	3.9969	3.0	1.0
4	24.0213	4.0	1.0
5	161.8142	50.0	1.0

Table 5.
Two variables example for a regular bad measurement.

$$X^{True} = [3, \ 12,]^{T}$$

It is clear to see that z_5 is a leverage point because of big 50.0.

3.2.1.1 WLS estimate

$$\hat{X}^{T} = [3.0805, \ 7.6000,] \ \Delta X^{T} = [-, 0.0805, \ 4.4,]$$

See **Table 6**.

i	z_i	$\hat{z}_i$	r_i
1	14.9954	10.6806	4.3149
2	18.0053	13.7611	4.2442
3	3.9969	16.8417	−12.8448
4	24.0213	19.9222	4.0991
5	161.8142	161.6275	0.1867

Table 6.
Results of WLS state estimator for two variables example for a regular bad measurement.

3.2.1.2 LAV estimate (selects z_1, z_5)

$$\hat{X}^T = [2.9963,\ 11.9991,]\ \Delta X^T = [0.0037\ 0.0009]$$

See **Table 7**.

i	z_i	$\hat{z}_i$	r_i
1	14.9954	14.9954	0.0
2	18.0053	17.9917	0.0135
3	3.9969	20.9880	−16.9912
4	24.0213	23.9843	0.0370
5	161.8142	161.8142	0.0

Table 7.
Results of LAV state estimator for two variables example for a regular bad measurement.

$$x_1^{true} = 3, x_2^{true} = 12$$

- The bad data has a significant negative influence on the WLS solution.
- The bad measurement was successfully rejected by LAV.

3.2.2 Test 2: Response to the leverage points

The above example shows that WLS state estimator is not able to reject normal bad measurements, while LAV state estimator is more robust than WLS state estimator because of being capable of rejecting bad measurements.

However, LAV is still suffering from the problem of leverage points. The definition of leverage points is given below:

A *leverage point* is a measurement that affects (dominates) the state estimator in such a way that the state estimate residuals will closely satisfy that particular measurement.

Influence of bad leverage point leads state estimators to a solution that is significantly biased due to the bad leverage point, and it is hard to detect those bad leverage points because their residuals are close to zero (**Table 8**).

z_5 is a bad leverage point. Standard deviation is 0.001. $x_1^{true} = 3, x_2^{true} = 12$.

i	z_i	A_{i1}	A_{i2}
1	14.9954	1.0	1.0
2	18.0053	2.0	1.0
3	20.9382	3.0	1.0
4	24.0213	4.0	1.0
5	20.0065	50.0	1.0

Table 8.
Two variables example for bad leverage point.

DOI: http://dx.doi.org/10.5772/intechopen.1012247

3.2.2.1 LAV estimate (selects z_1, z_5)

$$\hat{X}^T = \begin{bmatrix} 0.1023 & 14.8931 \end{bmatrix}$$
$$\Delta X^T = \begin{bmatrix} 2.8977 & -2.8931 \end{bmatrix}$$

See **Table 9**.

i	z_i	$\hat{z}_i$	r_i
1	14.9954	14.9954	0.0
2	18.0053	15.0977	2.9076
3	20.9382	15.1999	5.7383
4	24.0213	15.3022	8.7191
5	20.0065	20.065	0.0

Table 9.
Results of LAV state estimator for two variables example for bad leverage point.

3.2.2.2 SSE+ estimate (selects z_1, z_2)

$$\hat{X}^T = \begin{bmatrix} 3.0099 & 11.9855 \end{bmatrix}$$
$$\Delta X^T = \begin{bmatrix} -0.0099 & 0.0145 \end{bmatrix}$$

See **Table 10**.

Any combination of two measurements except z_5 is a solution of SSE+, which rejects the bad leverage point effectively.

i	z_i	***residual*** (z_1, z_2)	***residual*** (z_1, z_5)	***residual*** (z_2, z_4)	***residual*** (z_2, z_5)
1	14.9954	0.0	0.0	−0.0019	−2.9682
2	18.0053	0.0	2.9076	0.0	0.0
3	20.9382	−0.077	5.7383	−0.0751	2.8912
4	24.0213	−0.0038	8.7191	0.0	5.9326
5	20.0065	−142.474	0.0	−142.3828	0.0
# of abnormal residuals		3	3	1	3

Table 10.
Results of SSE+ state estimator for two variables example for bad leverage point.

4. The root causes of the vulnerability of traditional state estimators

This section explores the reasons why the performance of WLS and LAV estimators cannot meet the standards in real-world power system applications. These issues are not just implementation artifacts—they stem from fundamental assumptions baked into their mathematical formulations.

4.1 Including all measurements in the objective function vs. applying a rejection mechanism

In most traditional state estimation frameworks, like WLS state estimator, the optimization objective includes all available measurements, regardless of quality:t

$$J(x) = \sum_{i=1}^{m} w_i (z_i - h_i(x))^2$$

This approach implicitly assumes that the majority of measurements are correct and that any errors will "average out." However, even a few bad data points, especially if they are strategically located or highly weighted, can distort the entire solution.

A more robust alternative would apply a measurement selection or rejection mechanism prior to or during optimization. Rather than treating all data equally, the algorithm would aim to:

- Identify a subset of consistent measurements
- Exclude or downweight measurements likely to be erroneous.

This change in philosophy transforms state estimation from a global averaging problem into a local consistency evaluation, more aligned with practical grid behavior.

4.2 Evaluating residual magnitudes vs. counting abnormal residuals

Traditional state estimators, such as WLS and LAV, are fundamentally designed to minimize objective functions formed by residual magnitudes, where residuals are defined as the difference between measured and estimated values. Their objective functions, whether based on squared residuals (in WLS) or absolute residuals (in LAV), are built on the assumption that reducing the total residual magnitude leads to a more accurate estimate of the system state.

However, this reliance on residual magnitudes introduces a fundamental structural vulnerability. When a measurement is assigned a large weight, any deviation from its expected value contributes significantly to the objective function. To avoid a large penalty, the estimator is forced to reduce the residual of that measurement to near zero, even if the measurement is corrupted. This mechanism causes such measurements to become leverage points, exerting undue influence on the solution. If the leveraged measurement contains bad data, its residual still remains artificially small, effectively masking the error and allowing it to corrupt the overall state estimate.

Moreover, estimators that rely solely on residual magnitudes face additional limitations:

- Bad data at leverage points evade detection because their residuals are minimized by design.
- Model errors may distort residuals in misleading ways, leading to incorrect interpretations of measurement quality.

In contrast, SSE+ adopts a fundamentally different approach by shifting the objective from minimizing residual magnitude-based objective functions to minimizing the number of abnormal residuals. This reflects a shift in philosophy: rather than forcing every residual to be small, SSE+ allows some measurements to deviate if doing so improves the global consistency of the state estimate. By identifying and minimizing the count of residuals that exceed a consistency threshold, SSE+ is less vulnerable to leverage effects, better able to isolate bad data, and more structurally aware of how measurements interact. This leads to estimates that are not only numerically stable but also physically meaningful, even under stressed system conditions or when the model includes parameter or topology errors.

This transition from measuring "how much" a residual deviates to "how many" residuals are inconsistent represents a key innovation in state estimation and reflects the kind of structural robustness needed for future power system operations [9, 18].

4.3 Linearizing the objective function vs. preserving nonlinear characteristics

Traditional state estimators, including WLS and LAV methods, typically rely on linearizing the inherently nonlinear power flow equations. This is done to construct the Jacobian matrix required by iterative solvers such as Gauss-Newton. These nonlinear equations define the relationship between power injections, line flows, and bus voltages, but are approximated locally to enable tractable solution steps.

While linearization provides computational efficiency, particularly for large-scale systems, it introduces fundamental weaknesses that degrade both robustness and estimation quality:

- Ill-conditioned Jacobian matrix under stressed conditions: When the system is heavily loaded, near voltage collapse, or contains topology/parameter errors, the Jacobian matrix may become nearly singular or lose rank. This often leads to divergence or instability in Newton-type iterations.

- Strong dependence on the initial guess: Since linearization is only valid near the expansion point, a poor initial estimate can cause the algorithm to follow a misleading path, fail to converge, or settle on a physically implausible solution.

- Loss of critical nonlinear behaviors: Features like transformer tap changers, reactive power limits, and the behavior of lines with zero impedance are nonlinear in nature and cannot be captured accurately using first-order approximations. Linearized models tend to distort or suppress these effects.

- Propagation of bad data: The structure of the linearized system tends to diffuse the impact of bad measurements throughout the entire solution, making it difficult to isolate erroneous data without intensive post-processing.

In contrast, SSE+ avoids the linearization process entirely by solving the original nonlinear equations directly. It does not construct or depend on a Jacobian matrix, nor does it rely on step-by-step approximation techniques like Gauss-Newton. Instead, SSE+ retains the full nonlinear structure of the measurement model throughout the estimation process, enabling it to operate with greater numerical stability and higher physical fidelity.

Moreover, SSE+ adopts a sparsity-aware formulation that minimizes the number of abnormal residuals—rather than minimizing their magnitudes—while preserving nonlinear interactions between state variables and measurements. This provides several key advantages:

- Robust convergence even under extreme or misconfigured system conditions, since no fragile Jacobian matrix needs to be inverted or updated.
- Greater accuracy near the operating limits of the system, where linear approximations fail to capture system dynamics correctly.
- Natural treatment of non-standard components, such as tap transformers or partially observable buses, without artificial simplification.
- Reduced dependence on the initial state guess, as the estimation is guided by structural consistency in the measurement set, not by iterative local updates.

By working with the full nonlinear model from the outset, SSE+ shifts the trade-off between computational efficiency and physical accuracy. Rather than sacrificing realism for speed, SSE+ achieves both, using a structurally grounded formulation that scales efficiently and converges reliably—even under stress or uncertainty.

This highlights a crucial shift in the philosophy of state estimation: preserving the nonlinear nature of the problem is not a computational burden to be avoided, but a robustness-enhancing strategy essential for modern power systems [18].

4.4 Achieving a converged solution by adjusting measurement weights vs. obtaining voltage estimates best reflecting the operating state

Traditional WLS-based state estimators often achieve convergence by adjusting the weights of measurements during the iterative solution process. When faced with large residuals caused by bad data, topological or parameter errors, or stressed operating conditions, these estimators reduce the influence of problematic measurements—sometimes artificially—to force convergence. However, this practice can yield voltage estimates that do not accurately reflect the true operating state of the system. The convergence, in such cases, is numerical rather than physical, and the final state estimate may be significantly biased.

In contrast, SSE+ takes a fundamentally different approach. It focuses on identifying a clean basis set—a subset of measurements that is both sufficient for observability and minimally contaminated by bad data. Through compressive sensing and sparse recovery techniques, SSE+ effectively pushes bad or inconsistent measurements out of the basis set and into the redundant set. By solving the state estimation problem using only the clean basis set, SSE+ ensures that the resulting voltage estimates closely reflect the actual system conditions. This approach prioritizes physical accuracy over artificial convergence, yielding robust and reliable estimates even in the presence of significant measurement errors or modeling uncertainties [18].

5. Simulation results and performance evaluation

This section highlights the numerical results of state estimation under different conditions—ranging from ideal to highly corrupted data environments—focusing on convergence rate, bad data rejection, recovery accuracy, and computational efficiency [9].

5.1 Tests on 13659 bus system

The system data and measurement configuration for this test are given in **Table 11**.

In this study, three categories of bad data—topology errors, parameter errors, and measurement value errors—were systematically introduced into the system to evaluate estimator performance under realistic conditions. The measurement redundancy ratio, defined as the ratio of the total number of measurements to the number of state variables, was set at 2.51. This level reflects a moderate redundancy typical in practical applications. For each type of bad data, the number of erroneous entries varied from 10 to 500, and 100 randomized trials were conducted for each case. To further assess the impact of significant errors, topology and parameter faults were injected specifically on branches carrying at least 80 MW of real power flow. Across all test scenarios, the SSE+ algorithm demonstrated a 100% convergence rate. The results for all three types of bad data are summarized in **Figures 5–7**.

5.2 Test on a real power system case

SSE+ was tested on about 100 real-world cases provided by four major utility companies in North America. This subsection presents the testing results for one representative case. Detailed system information and the corresponding measurement configuration are summarized in **Table 12**. Notably, the measurement redundancy ratio in this case is relatively low, reflecting the challenges typically encountered in practical utility environments.

We tested the solution quality and capability of bad data recovery for SE+ in the following three aspects.

5.2.1 Comparison of voltage solution quality with a standard commercial SSE+

For this real-world system case, the utility also provided the voltage solution obtained using their standard commercial state estimator. To evaluate and compare the quality of voltage estimates produced by SSE+ and the commercial SSE+, we computed the residuals of the voltage measurements for both solutions. The comparative results are presented in **Table 13**.

Voltage measurement residuals are defined as the differences between the measured bus voltage magnitudes and the corresponding solved voltage magnitudes. Smaller residuals indicate that the estimated voltages closely match the actual measurements. To facilitate an overall comparison, we sorted the voltage residuals from both solutions and plotted them as two curves on the same graph. As shown in

	Configuration
Total # of buses	13659
Total # of branches	20467
Total # of line flow measurements	35510
Total # of bus injection measurements	27318
Total # of bus voltage magnitude measurements	5772
Measurement redundancy ratio	2.51

Table 11.
Measurement configuration of 13659 bus system.

Figure 5.
Rejection testing on the typical three bad data.

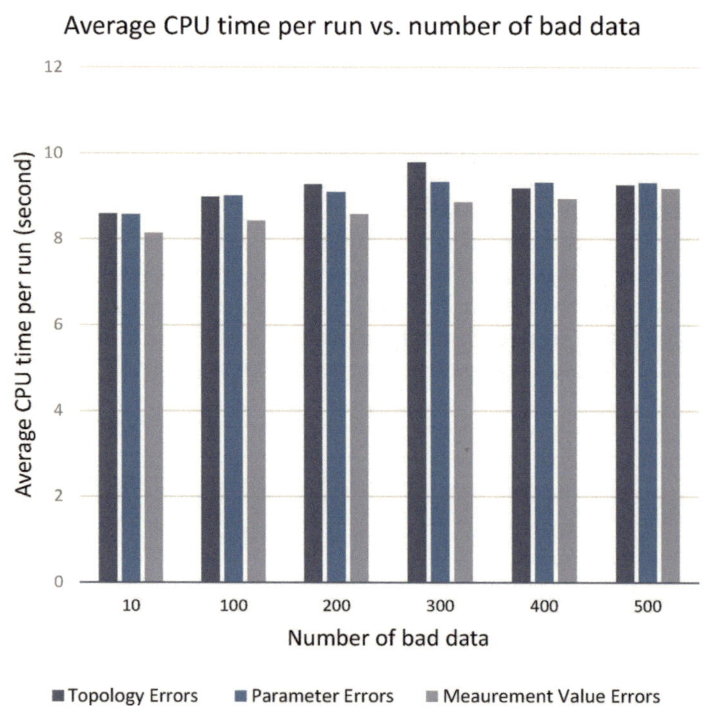

Figure 6.
Calculation time testing on typical three bad data. The PC computer has 4 cores with 1.80GHz.

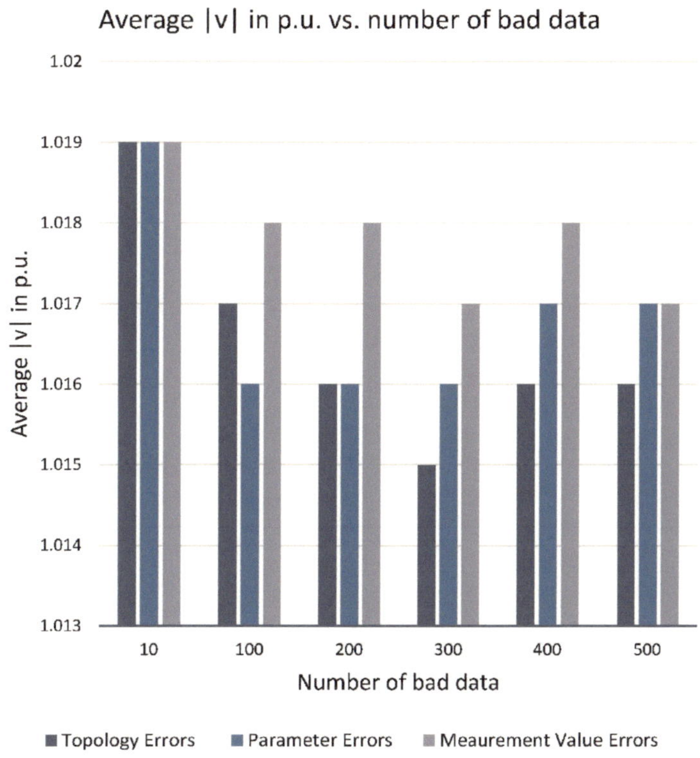

Figure 7.
Average voltage magnitudes for the testing of typical three bad data.

# of buses	3409
# of transmission branches	4337
# of line flow measurements	3571
# of injection measurements	3022
# of voltage measurements	1457
Measurement Redundancy ratio	2.15

Table 12.
System data and measurement configuration of the utility real case.

	Standard commercial SSE+	SSE+
Sum of voltage measurement residuals	15.45 p.u.	12.97 p.u.
Average of voltage measurement residuals	0.0108 p.u.	0.0091 p.u.
Standard deviation	0.0163	0.0158
Mean voltage solution value in p.u.	1.0259	1.0257

Table 13.
Comparison of voltage solution quality.

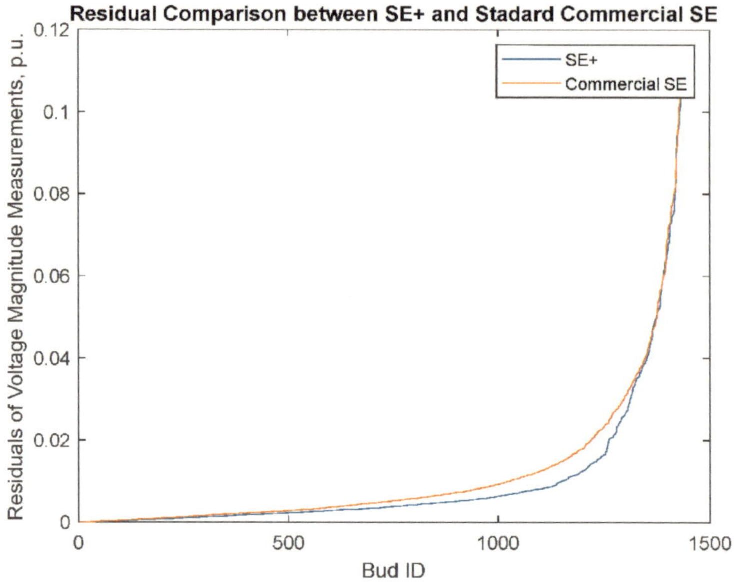

Figure 8.
The comparison of voltage residuals between solutions of commercial SSE+ and SSE+.

Figure 8, the voltage estimates produced by SSE+ consistently exhibit smaller residuals, demonstrating that SSE+ provides a solution more closely aligned with the measured data than the standard commercial SSE+.

5.2.2 Recovery of accurate values lost due to measurement errors

It is well known that the WLS state estimator is generally unable to recover the true values lost due to bad measurements, as it lacks the inherent capability to reject erroneous data. In contrast, the SSE+ approach effectively mitigates the impact of bad measurements by constructing a base set composed of the minimum number of most consistent measurements. This process naturally excludes inconsistent—or bad—measurements from the estimation, thereby enhancing the reliability of the solution. As a result, the voltage estimates produced by SSE+ are not affected by rejected bad data and are expected to be more accurate. To evaluate the bad measurement recovery capability of SSE+, we conducted the following two tests.

5.2.2.1 Recovery of bad data on injection measurements

In this test, large errors of 1000 MW were intentionally introduced to 10 real power injection measurements. Due to the low measurement redundancy ratio in the system, SSE+ selected five of these ten measurements to form part of the base set, while the remaining five were placed in the redundant set. Remarkably, four of the bad measurements were accurately recovered by SSE+, as illustrated in **Figure 9**.

One of the bad injection measurements could not be accurately recovered: the original value was 0.0 MW, while the recovered value was 3424.0 MW. This large deviation occurred because the voltage at the associated bus was biased, caused by a neighboring bad injection measurement that was included in the base set. However,

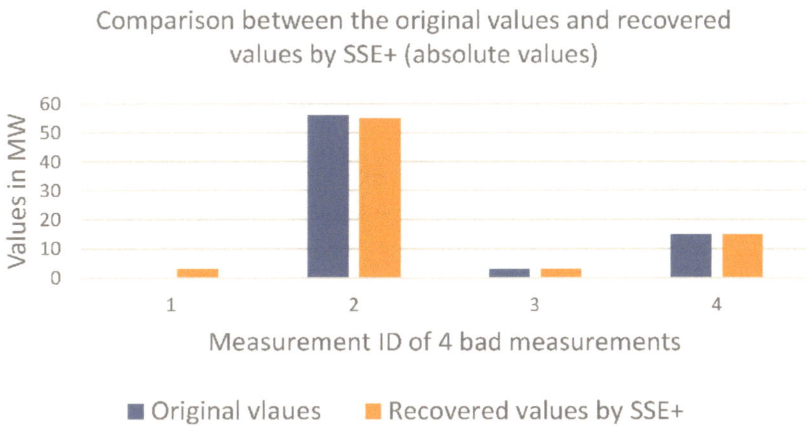

Figure 9.
Test results for bad data recovery for bus injection measurements.

the fact that the other four bad measurements were precisely recovered indicates that the negative impact of bad measurements within the base set is limited and localized. SSE+ effectively isolates and minimizes the influence of such errors, enabling accurate recovery in most cases even under challenging conditions.

5.2.2.2 Recovery of bad data on powerline flow measurements

To evaluate the recovery capability of SSE+ for contaminated line flow measurements, we introduced large errors of 500 MW to 10 real power line flow measurements. Among these, SSE+ selected one bad measurement to enter the base set, while the remaining nine were rejected and placed in the redundant set. To illustrate the recovery accuracy, **Figure 10** presents the differences between the original values and their recovered counterparts. The small differences shown in the figure indicate that

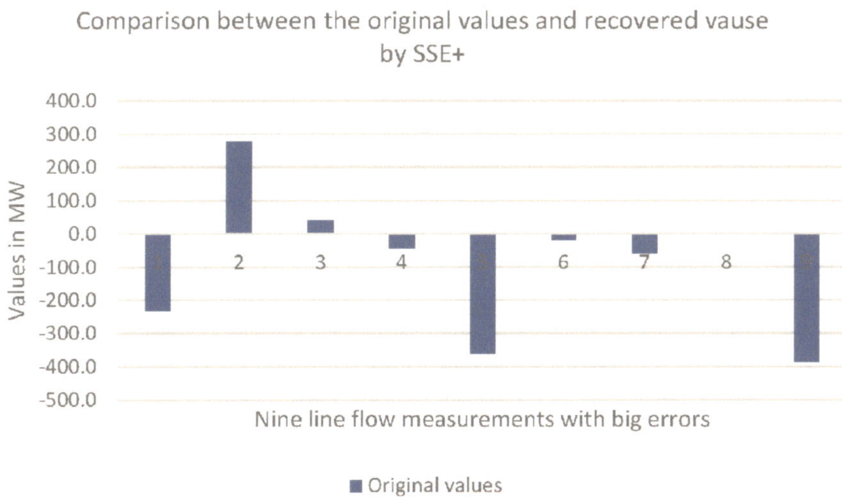

Figure 10.
Testing of bad data recovery for the line flow measurements: big errors on line flow measurements.

SSE+ accurately recovered the contaminated line flow measurements, even when some erroneous data were present in the base set (**Figure 11**).

5.3 Computational efficiency and scalability

The computational performance of SSE+ was evaluated using a diverse set of real-world power systems of varying sizes, as illustrated in **Figure 12**. The number of buses in these systems ranges from approximately 2500 to over 13,000. Despite the

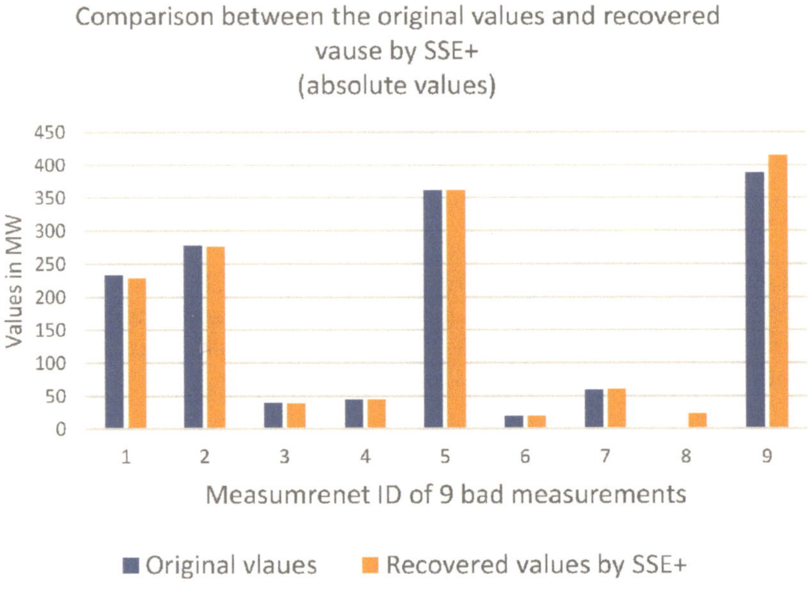

Figure 11.
Testing of bad data recovery for the line flow measurements: recovered values of line flow measurements with big errors.

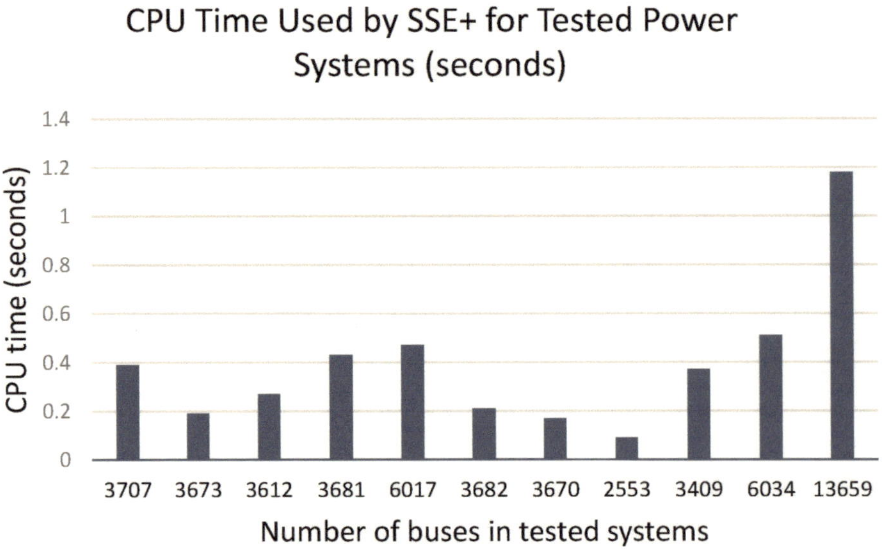

Figure 12.
The CPU time of SSE+ for power systems with different sizes.

significant variation in system size and complexity, SSE+ consistently demonstrates high computational efficiency. For all systems containing fewer than 7000 buses, the state estimation process is completed in under 0.6 seconds. Even in the largest case—comprising 13,659 buses—SSE+ required only about 1.2 seconds of CPU time. These results underscore the scalability and real-time capability of the SSE+ algorithm. Its ability to deliver robust and accurate state estimates with minimal computational burden confirms its suitability for deployment in modern energy management systems.

6. Conclusions and future work

This chapter aimed to revisit the fundamentals of power system state estimation, focusing on basic concepts, practical challenges, and small illustrative examples, rather than on advanced mathematical theories or abstract generalizations. The intent was to provide a clear and accessible understanding of how traditional estimators operate, why they often fail under real-world conditions, and what conceptual shifts are needed to develop more robust and reliable estimation methods for modern power systems.

The analysis centered primarily on the two most widely used estimation techniques in power systems: Weighted Least Squares (WLS) and Least Absolute Value (LAV) estimators. These methods are grounded in general-purpose optimization frameworks that assume measurement noise is Gaussian, the system model is accurate, and the estimation problem is well-conditioned. However, this chapter demonstrated that these assumptions break down frequently in actual grid operations—where measurement errors, topology inaccuracies, parameter uncertainties, and stressed system conditions are common.

Through simplified examples and structural analysis, the chapter revealed the deep underlying causes of the limitations inherent in these estimators. WLS, in particular, distributes the influence of all measurements—accurate or flawed—across the entire system state vector. When even a single bad data point exists, especially at a leverage point, it can bias the solution without being detected by standard residual analysis. In stressed conditions, such as after disturbances or during high loading scenarios, the linearized Jacobian matrix used in Newton-based solvers may become ill-conditioned, leading to poor convergence or complete divergence. These failures are not anomalies but rather consequences of core design assumptions that treat all data equally and depend on residual magnitudes as the primary signal of inconsistency.

To illustrate these issues, a simple 4-bus example was used to show how traditional estimators behave in the presence of bad data. Even in this minimal setting, WLS SE failed to produce reliable results when incorrect measurements were introduced. These examples were deliberately kept small in order to isolate and visualize the behavior of the algorithms, making it easier to trace the source of failure and to expose the conceptual weaknesses of the traditional estimation framework.

In contrast, the chapter introduced SSE+, a more robust estimator designed with the unique characteristics of power systems in mind. Unlike general-purpose estimators, SSE+ does not attempt to fit all data equally. Instead, it operates on the principle of selectivity and consistency, identifying a minimal subset of measurements that are physically consistent and trustworthy, and solving the estimation problem primarily from this clean basis set. This shift in philosophy—away from full-data optimization and toward structure-aware selection—represents a fundamental departure from traditional methods. SSE+ is not designed for arbitrary systems, but rather for the specific challenges posed by real-world power grids, including limited observability, network nonlinearities, and error-prone measurements.

The examples presented in the chapter demonstrate this philosophical difference. While WLS aims to minimize global error through averaging, SSE+ seeks to recover the most physically meaningful and internally consistent solution, even if that means discarding a portion of the available data. This makes SSE+ inherently more robust to bad data, less dependent on ideal assumptions, and more reflective of actual system behavior.

These insights have important implications for power system operations. As power grids become more complex, with increasing integration of renewables, growing measurement diversity, more frequent extreme weather events, and heightened cybersecurity risks, the need for reliable and adaptable estimation grows. Traditional residual-based bad data detection techniques are inadequate in the presence of topological uncertainty, correlated errors, or stealthy data manipulation. Estimators must evolve to incorporate mechanisms that can detect inconsistency patterns, adapt to modeling inaccuracies, and respond intelligently to conflicting information.

Looking ahead, the future of state estimation lies in developing algorithms that are selective, resilient, and system-aware. One promising direction is the advancement of data-selective estimators—methods that focus on identifying a self-consistent core of data and solving the estimation problem from that subset, rather than attempting to reconcile all measurements. This approach directly addresses the limitations exposed in this chapter and aligns more closely with how real operators evaluate system reliability: not by assuming everything is correct, but by identifying what can be trusted.

Another critical area is the integration of machine learning and artificial intelligence into the estimation process. These tools can help identify suspicious measurements, adjust weights dynamically, or learn system behavior patterns that may indicate model errors or cyberattacks. When combined with physical modeling, AI-driven insights can enhance both accuracy and robustness.

Additionally, hybrid estimation frameworks that fuse data from SCADA and PMUs can increase observability and redundancy, offering more reliable estimates even under stressed or abnormal operating conditions. As measurement infrastructure evolves, estimators must be capable of seamlessly integrating diverse data sources with different resolutions, time scales, and error characteristics.

Moreover, in a world where cyber-physical threats are increasingly sophisticated, the development of cybersecurity-aware estimation techniques is essential. Estimators must be designed not only to tolerate noise but also to identify and reject intentional data manipulation, stealth topology alterations, and coordinated attacks that traditional techniques cannot detect.

Finally, there is a growing need for formally robust estimation algorithms that offer performance guarantees under specified conditions. Estimators with bounded error characteristics and provable resilience to data corruption will become essential tools in mission-critical operations and regulatory compliance.

Author details

Bei Gou[1]*, Qing Zhao[2] and David Bogen[2]

1 Grid Science Solutions, LLC, Mansfield, TX, USA

2 Oncor Electric Delivery Inc., TX, USA

*Address all correspondence to: bgou@smart-eleecgrid.com

References

[1] York State Energy Research and Development Authority (NYSERDA). Fundamental research challenges for distribution state estimation to enable high-performing grids. In: NYSERDA Report Number 18–37. Prepared by Smarter Grid Solutions. New York, NY: New York State Energy Research and Development Authority (NYSERDA); 2018. nyserda.ny.gov/publications

[2] Schweppe FC, Wildes J. Power system static-state estimation, part I: Exact model. IEEE Transactions on Power Apparatus and Systems. 1970; **PAS-89**(1):120-125. DOI: 10.1109/TPAS.1970.292678

[3] Meliopoulos APS, Fardanesh B, Zelingher S. Power system state estimation: modeling error effects and impact on system operation. In: Proceedings of the 34th Annual Hawaii International Conference on System Sciences, (HICSS), Maui, HI, USA; 2001. pp. 682-690. DOI: 10.1109/HICSS.2001.926269

[4] Schweppe FC, Rom DB. Power system static-state estimation, part II: Approximate model. IEEE Transactions on Power Apparatus and Systems. 1970; **PAS-89**(1):125-130. DOI: 10.1109/TPAS.1970.292679

[5] Schweppe FC. Power system static-state estimation, part III: Implementation. IEEE Transactions on Power Apparatus and Systems. 1970; **PAS-89**(1):130-135. DOI: 10.1109/TPAS.1970.292680

[6] Cibaku E, Gama F, Park SW. Boosting Efficiency in State Estimation of Power Systems by Leveraging Attention Mechanism. Faculty Publications; 2024. p. 457. Available from: https://digitalcommons.njit.edu/fac_pubs/457

[7] Federal Energy Regulatory Commission (FERC), North American Electric Reliability Corporation (NERC), and Regional Entity Staff. The February 2021 cold weather outages in Texas and the South Central United States. 2021. FERC (.gov)

[8] Ji Y, Kim J, Thomas RJ, Tong L. Forecasting real-time locational marginal price: A state space approach. In: Proceedings of the Asilomar Conference Signals, Systems and Computers (ACSSC), Pacific Grove, CA, USA. 2013. pp. 379-383. DOI: 10.1109/ACSSC.2013.6810300

[9] Gou B, Shue D. Advances in algorithms for power system static state estimators: An improved solution for bad data management and state estimator convergence. IEEE Power and Energy Magazine. 2023;**21**(1):16-25. DOI: 10.1109/MPE.2022.3219168

[10] Merrill HM, Schweppe FC. Bad data suppression in power system static state estimation. IEEE Transactions on Power Apparatus and Systems. 1971;**PAS-90**(6):2718-2725. DOI: 10.1109/TPAS.1971.292925

[11] Chen Y, Zhou N. A comparative study on state estimation algorithms for power systems. In: Proceedings of the 2020 52nd North American Power Symposium (NAPS), Tempe, AZ, USA. 2021. pp. 1-6. DOI: 10.1109/NAPS50074.2021.9449766

[12] Dy Liacco TE. The role of state estimation in power system operation. IFAC Proc. 1982;**15**(4):1531-1533

[13] Mili L, Phaniraj V, Rousseeuw PJ. Least median of squares estimation in power systems. IEEE Transactions on Power Systems. 1991;**6**(2):511-523. DOI: 10.1109/59.76693

[14] Zhao J, Mili L, Pires R. Statistical and numerical robust state estimator for heavily loaded power systems. IEEE Transactions on Power Systems. 2018; **33**(6):6904-6914

[15] Chen J, Liao Y, Gou B, Yocum K. Impacts of load levels and topology errors on WLS state estimation convergence. In: Proceedings of the 2012 North American Power Symposium (NAPS), Champaign, IL, USA. 2012. pp. 1-6. DOI: 10.1109/NAPS.2012.6336329

[16] MIT Energy Initiative, The Future of the Electric Grid: An Interdisciplinary MIT Study. Cambridge, MA, USA: Massachusetts Institute of Technology, 2011. [Online]. Available: https://energy.mit.edu/wp-content/uploads/2011/12/MITEI-The-Future-of-the-Electric-Grid.pdf. ISBN: 978-0-9828008-6-7

[17] Abur A, Gómez-Expósito A. Power System State Estimation: Theory and Implementation. New York, NY, USA: Marcel Dekker; 2004

[18] Robust SE. Home [Internet]. Smart-elecgrid.com. 2023. Available from: https://www.smart-elecgrid.com/

Section 3

Renewable Energy Systems

Chapter 4

Solar Power Plant (Solar PV) Technology, Industry, Local Content, and Risk Mitigation: Case Study in Indonesia

Agus Krisnowo, Oo Abdul Rosyid, Iwan Sudrajat, Jaizuluddin Mahmud, Gani Soehadi, Ignatius Subagjo, Irawan Santoso, Mohamad Dahsyat, Yustiar Gunawan, Syaeful Karim, I Made Astawa, Ugay Sugarmansyah and Susalit Setya Wibowo

Abstract

Indonesia, located on the equator, receives consistent sunlight exposure throughout the year with an average duration of 10–12 hours per day. Therefore, solar PVs, as a new and renewable energy source, are a primary focus for the future. Solar PV is a power-plant that utilizes the photovoltaic effect from direct sunlight energy. The principle of photovoltaics is to convert photon energy from sunlight into electrical energy. Silicon-based solar modules and bifacial solar modules are the most dominant technologies in Indonesia, both for large-scale solar PVs such as floating and ground-mounted solar PV, rooftop solar PV, and off-grid systems in remote areas. Although solar PVs have been built in several countries, including Indonesia, efforts to improve technology, industry, local content, and risk mitigation remain necessary. This relates to efforts to increase solar cell efficiency, enhance domestic industrial capabilities, reduce costs, and minimize risks that arise during the construction process.

Keywords: solar PV, technology, industry, local content, risk mitigation, efficiency, reduce cost

1. Introduction

In recent decades, awareness of the use of renewable energy in all countries has been increasing, particularly due to concerns about the depleting availability of fossil energy resources and the occurrence of climate change. This awareness has emerged not only from governments but also from the user community. One of the renewable and future energy sources is solar PV that utilizes the photovoltaic effect from direct sunlight energy. According to the International Energy Agency (IEA) in 2024, solar PV experienced a growth of 475 TWh, representing a 30% increase, marking the largest growth among all power generation technologies to date. This chapter will discuss solar PV as a new and renewable energy source for the future and its development in Indonesia, covering aspects of technology, industry, local content, and risk mitigation.

Generally, the components of a solar PV consist of solar modules, inverters, frames, distribution panels, transformers, AC/DC cables, protection systems, and DC combiner boxes, as shown in **Figure 1**.

The development of technology related to solar PV includes various innovations to improve efficiency, reduce costs, and expand applications. Key developments include high-performance solar panel technology, transparent solar panels, improved energy storage systems, and integration with the Internet of things (IoT).

The Indonesian government, through Government Regulation No. 79/2014 on the National Energy Policy (KEN), has set a target for new and renewable energy (EBT) to contribute 23% to the national energy mix by 2025 [1]. This regulation was followed by Presidential Regulation No. 22/2017 on the National Energy General Plan (RUEN), which details the utilization targets for each type of renewable energy [2]. Additionally, Indonesia ratified the Paris Agreement in December 2015 (COP 21) through a law aimed at reducing greenhouse gas (GHG) emissions, committing to a 29% reduction by 2030 with its own efforts and up to 41% with international support [3]. The main component of solar PV is solar module, which converts solar energy into electricity. Solar PV is an environmentally friendly energy source that produces no greenhouse gas emissions during operation. This technology is becoming an increasingly important choice for the global community in reducing dependence on fossil fuels. Initially, the Indonesian government targeted an installed solar energy capacity of 6.5 GW by 2025 [2]. However, due to various factors, this target was

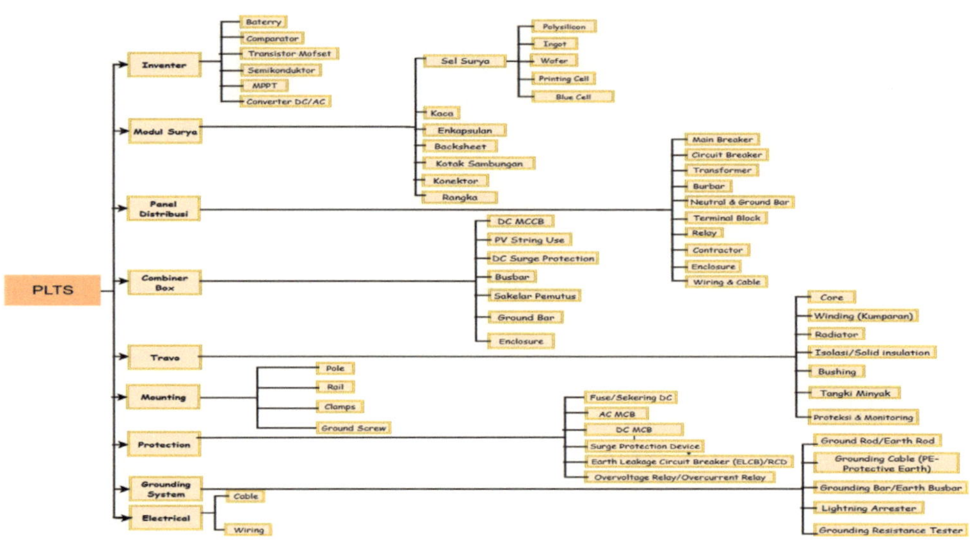

Figure 1.
Solar PV industrial tree.

reduced to around 870 MW [4]. According to the Ministry of Energy and Mineral Resources (ESDM), as of 2023, the installed capacity of solar PV in Indonesia reached 573.8 MW [5]. One effort to boost solar PV adoption is the regulation of rooftop solar PV. By March 2025, the government recorded an installed rooftop solar PV capacity of 406.78 MW, consisting of 10,437 PLN customers [6].

The rapid growth of rooftop solar PV is driven by regulations governing its installation. There are five legal frameworks regulating rooftop solar PV: [7] Regulation of the Minister of Energy and Mineral Resources No. 2/2024 on Rooftop Solar PV Connected to the Electricity Network, Government Regulation No. 26/2021 on the Implementation of the Energy and Mineral Resources Sector, Regulation of the Minister of ESDM No. 12/2019 on the Capacity of Power Plants for Self-Use Based on Operating Permits, Regulation of the Minister of ESDM No. 49/2018 on the Use of Rooftop Solar PV by PLN Consumers, and Government Regulation No. 14/2012 on Electricity Supply Business Activities. In addition to rooftop solar PV, another strategy is to replace or integrate existing diesel powerplants scattered across various regions, particularly in isolated networks, with renewable energy plants, including solar PV, by considering technical and economic feasibility. The utilization of renewable energy plants depends on the potential of each region in Indonesia, with solar PV being one of the options.

The main objective of this chapter is to explore advancements in solar PV technology, encompassing solar modules, inverters, and batteries, and their implementation in Indonesia. This includes the development of the industry and its local components, as well as risk management in solar powerplants.

2. Technology of solar PV

This section provides an in-depth look at the ongoing technological innovations in solar power generation systems, focusing on advancements in solar modules, energy storage, and inverter technologies, with an emphasis on improving efficiency, reducing costs, and ensuring reliable energy storage and conversion.

The principle of solar PV is to convert photon energy from sunlight into electrical energy. This conversion occurs in solar cells, which consist of thin layers of pure silicon (Si) and other semiconductor materials. When these materials receive photon energy, electrons are released from their atomic bonds, becoming free-moving electrons that generate direct current (DC) voltage. The photovoltaic effect was first discovered by Edmund Becquerel and was used in the satellite industry in the 1950s. Subsequently, in the 1960s, photovoltaics began to be applied in various applications [8].

The schematic working principle of a solar cell is shown in **Figure 2**. The electric current produced by a solar cell exposed to sunlight results from the absorption of photons, which generate electron–hole pairs. These holes and electrons are then collected and separated by a p-n junction, creating a potential difference between the p-type and n-type layers. This results in an open-circuit voltage (Voc) across the cell's terminals. When the top and bottom contacts of the solar cell are connected to a load via electrical cables, an electric current is produced. To maximize the power output of the solar cell, it must be designed to absorb sunlight as efficiently as possible. A collection of photovoltaic cells connected in series, parallel, or a combination of both forms a photovoltaic module/solar module.

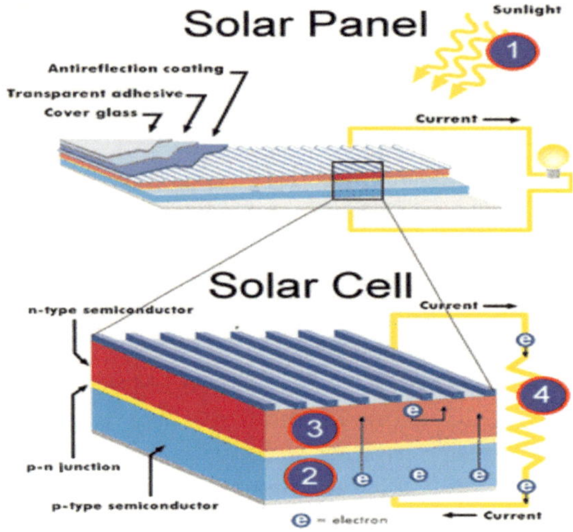

Figure 2.
Principle of a solar cell/photovoltaic [9].

2.1 Development of solar module technology

The development of solar PV technology focuses on improving the efficiency of solar modules. This is driven by the need for efficient, cost-effective, and environmentally friendly renewable energy. Below is an overview of the latest trends and innovations in solar module technology.

a. *Increased solar module efficiency*

- *Silicon solar cells*: Monocrystalline and polycrystalline silicon technologies continue to dominate the market due to their balance of efficiency and cost. Monocrystalline silicon-based solar modules now achieve commercial efficiencies of 22–24%, with technologies like passivated emitter and rear cell (PERC) and half-cut cells increasing power output. PERC is a high-efficiency solar cell architecture distinct from conventional monocrystalline cells. **Figure 3** illustrates the structural differences between conventional Al-BSF cells and

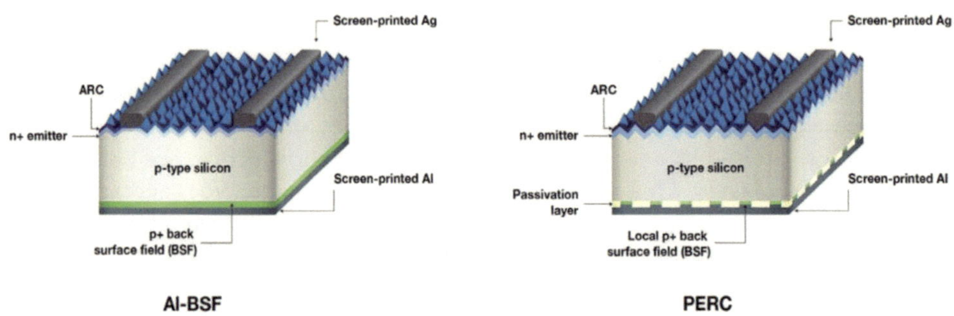

Figure 3.
Illustrates the structural differences between conventional Al-BSF cells and PERC cells [10].

PERC cells [10]. Another innovation is the half-cut cell technology, with REC introducing the first commercial half-cut cells in 2014 [10].

- *Next-generation solar cells*: Silicon-perovskite tandem solar cells show significant potential, with efficiencies ranging from 26 to 30% [11, 12]. Laboratory tests have even exceeded 30%, although stability remains a research focus [13]. These advancements could lead to more efficient and cost-effective solar energy solutions. The working principle of perovskite-silicon tandem photovoltaics leverages the strengths of both materials. Perovskite efficiently absorbs high-energy photons, while silicon is better at absorbing low-energy photons. By combining these layers, the tandem structure captures a broader spectrum of sunlight, improving energy conversion efficiency and power generation.

b. *Bifacial technology*

- Bifacial solar modules, which absorb light from both sides (front and back), are gaining popularity due to their ability to increase energy yield by 10–20% depending on environmental conditions. Innovations in anti-reflective coatings and rear surface designs have enhanced the performance of bifacial solar modules. Bifacial solar modules do not use a white backsheet but instead use a transparent layer (glass). Bifacial technology offers several advantages, including reduced microcracking, delamination, and moisture corrosion, lower cell temperatures, no potential-induced degradation due to the absence of a metal frame requiring grounding, lower degradation rates, higher fire resistance, and high mechanical strength with minimal flexing. **Figure 4** illustrates the working system of bifacial solar modules.

c. *Alternative materials and thin-film*

- Thin-film solar cells, such as copper indium gallium selenide (CIGS) and CdTe, are being developed for specific applications like building-integrated photovoltaics (BIPV). These technologies offer flexibility and lightweight properties, although their efficiency (14–18%) is still lower than silicon [15].

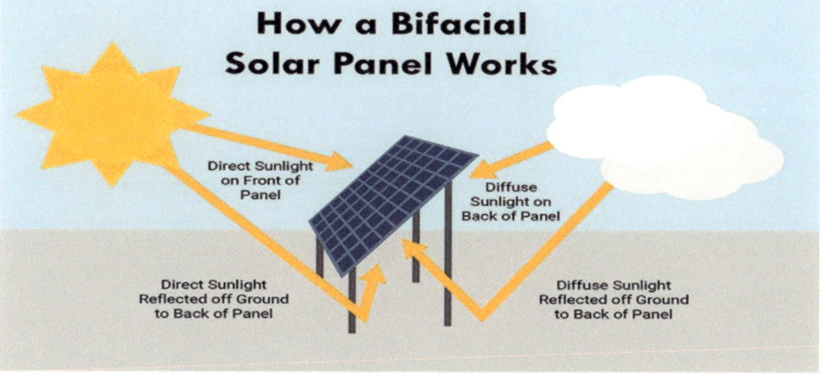

Figure 4.
Illustrates the working system of bifacial solar modules [14].

- Perovskite has garnered attention due to its low production costs and adaptability to flexible surfaces [11–13].

d. *Integrasi Teknologi Digital dan Smart PV*

- Modern solar modules are increasingly integrated with digital technologies, such as the Internet of things (IoT) and artificial intelligence (AI), for performance optimization. AI-based monitoring systems are used to detect module damage and maximize energy output [16].

- Maximum powerpoint tracking (MPPT) technology with specific methods can improve system efficiency under partial shading conditions [17].

e. *Sustainability and recycling*

- The focus on sustainability has driven research into recycling solar modules and using environmentally friendly materials. Recycling processes for silicon, glass, and metals from used modules have been improved to reduce waste [18–20].

2.2 Development of energy storage and management technology

In solar PV systems, batteries are essential for managing the variability of fluctuating renewable energy generation. Batteries stabilize power output, whether used standalone or in conjunction with other powerplants. Battery storage technology for solar PV has seen significant advancements, driven by the need to enhance efficiency, reliability, and sustainability of renewable energy. Various battery technologies are used in solar PV systems, including lithium-ion, lead-acid, and gel-cell batteries. Lithium-ion is the preferred choice due to its high efficiency, longer lifespan, and greater storage capacity. A BNEF survey shows that the cost of lithium-ion batteries has fallen 90% since 2010 and is expected to reach below $100/kWh by 2030, as shown in **Figure 5**. This will make solar PVs with storage increasingly competitive [21].

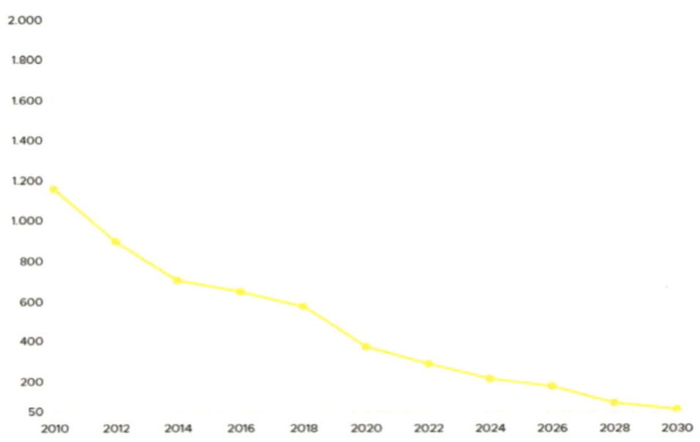

Figure 5.
Global lithium battery cost decline based on BNEF survey [21].

2.3 Development of inverter technology

Inverters are electrical devices used to convert direct current (DC) into alternating current (AC). Inverters convert DC from devices such as batteries or solar cells into AC. There are two types of inverters: solar inverters (on-grid) and battery inverters (off-grid). Both inverters must be connectable via an AC electrical network. This allows communication between solar inverters and battery inverters located far apart. With this feature, all inverters can communicate using only the AC power line without requiring additional communication networks (**Figure 6**).

The use of inverters in solar PV is to supply electricity to devices that use alternating current (AC). Several factors to consider when selecting an inverter include load capacity in watts. The inverter capacity must be based on the load to be used, with input DC voltages of 12 V, 24 V, or 48 V and output AC types of sine wave or square wave.

Pure sine wave inverter technology is considered the best currently available. The waveform produced by a pure sine wave inverter is nearly identical or even better than the electricity supplied directly by PLN. The total harmonic distortion (THD) of a pure sine wave inverter is less than 3%, making it ideal for all electronic devices. Pure sine wave inverters produce AC waveforms with low THD (1.6% in simulation and 0.05% in testing with cuckoo optimization algorithms), making them the best choice for applications requiring clean and stable power, such as solar PV [23, 24]. Consequently, pure sine wave inverters are the most expensive due to their output waveform closely resembling the original grid waveform. In the market, modified sine wave inverters, a combination of square wave and sine wave, are also available. When viewed through an oscilloscope, the waveform appears as a sine wave with dashed lines between the y = 0 axis and the sine graph. Devices with coils can still operate with modified sine wave inverters, though not optimally. Square wave inverters, however, cannot operate loads with coils or motors at all.

Additionally, grid tie inverters are specialized inverters typically used in renewable energy systems, converting DC electricity into AC that is fed into the existing electrical grid. Grid tie inverters, also known as synchronous inverters, synchronize the generated AC current with the frequency and phase of the existing grid, ensuring stable and safe electricity flow [25]. With grid tie inverters, excess kWh generated by the solar PV system can be supplied to the PLN grid for shared use, with PLN compensating the solar PV provider at a pre-agreed tariff.

Losses in inverters typically occur as power dissipation in the form of heat. Grid tie inverters claim the highest efficiency, reaching 95–97% when the output load is close to its rated capacity [26, 27]. Generally, inverter efficiency ranges from 50 to 90%,

Figure 6.
Inverter [22].

depending on the output load. The closer the output load is to the inverter's rated capacity, the higher the efficiency, and vice versa. Modified sine wave or square wave inverters, when used for inductive loads, have significantly lower efficiency compared to pure sine wave inverters, consuming 20% more power than necessary.

3. Connecting solar PV to the grid

This section focuses on the integration of solar powerplants with the electrical grid, discussing different system configurations and the challenges of ensuring reliable power supply. The electricity from solar PV connected to the grid requires an interactive system between the solar PV and the electrical grid. Grid interconnection requires a system or device known as power conditioning. Power conditioning functions to maximize PV power output and convert DC voltage into high-quality AC voltage that meets the electrical parameters of the grid. In conventional powerplants, the power or energy generated can be controlled based on fuel supply. For solar PV, the energy generated is more variable, depending on the solar radiation received by the solar modules. Therefore, solar PV cannot operate independently and requires support from a dynamic grid system to maintain electrical system stability.

The integration of solar PV into the grid reduces the capacity or operation of conventional powerplants, thereby reducing fuel consumption and emissions. However, the reduced capacity of conventional powerplants due to solar PV electricity supply creates other challenges, particularly for coal-fired powerplants. Coal-fired powerplants lack the ability to quickly ramp up or down their capacity. Solar PV is more suitable for integration with diesel powerplants due to their ability to respond quickly to load changes, which is crucial for maintaining system stability. However, the solar PV capacity connected to the system must not exceed 20% of the peak load. If the solar PV capacity exceeds 20% of the peak load (during the day), it surpasses the system's flexibility limit, and the energy generated by solar PV will be rejected to maintain system stability. The system topology represents the design basis of the system, describing the relationship between solar PV, the load to be supplied, and the existing electrical grid (connected to other powerplants). All analyzes of the system topology will guide the determination of component requirements and interactions within the solar PV, such as the total PV power capacity, inverters, charge controllers, and batteries. Generally, based on their connection to the electrical grid, there are three system topologies: on-grid, off-grid, and hybrid.

3.1 On-grid

Grid-connected or on-grid solar PV systems are connected to the existing electrical grid (e.g., PLN). Thus, the system works in conjunction with the grid to meet load requirements. When sunlight is below the minimum operating threshold, the electrical demand can still be met by the grid. The main concepts of on-grid solar PV are as follows:

a. Most systems do not use batteries, making the system simpler and more affordable.

b. Reduces transmission power losses due to the proximity of solar PV to the load.

c. Acts as a backup power source.

However, in certain large-scale solar PV systems, the absence of batteries can have negative impacts. Fluctuations in solar PV power production (depending on solar radiation) can cause voltage fluctuations in the system. This can be mitigated by installing batteries to stabilize grid voltage (not as large-scale energy storage for load withdrawal outside sunlight hours). The schematic of an on-grid solar PV system is shown in **Figure 7**.

3.2 Off-grid

Off-grid systems are power generation systems that rely solely on solar energy as the primary energy source, using photovoltaic module arrays to generate electricity. This system is a suitable alternative for areas difficult to reach by large-scale power grids like PLN or large-scale diesel powerplants. Since they are not supported by a large electrical grid, battery energy storage systems (BESS) are critical. Additionally, an alternative to improving system reliability is adding other power generation sources to the network, such as generators, solar PV, micro-hydro, or wind power. The schematic of an off-grid solar PV system is shown in **Figure 8**.

The concept of a hybrid off-grid solar PV is to replace diesel powerplants with a lifespan exceeding 15 years, which typically have high specific fuel consumption (SFC), with solar PV. Solar PV operates only during sunlight hours, typically from 7:00 AM to 5:00 PM on clear days, while batteries supply power at night. In this system, the battery inverter acts as a grid-former (voltage source) as a voltage reference. Energy generated by PV is stored in the battery when PV production exceeds the load, as long as the battery's state of charge (SoC) has not reached its maximum. SoC indicates the level of energy stored in the battery. Maximum SoC indicates a fully charged battery, while minimum SoC indicates the battery can no longer supply energy.

When the battery is fully charged, it enters a floating mode, allowing it to remain a grid-former, and PV output power must be reduced to match the load requirement (curtailing). When PV power is less than the load demand (e.g., in the evening, at night, or during cloudy conditions), the battery supplies power to the load as long as the SoC remains above the minimum threshold. This system does not require external

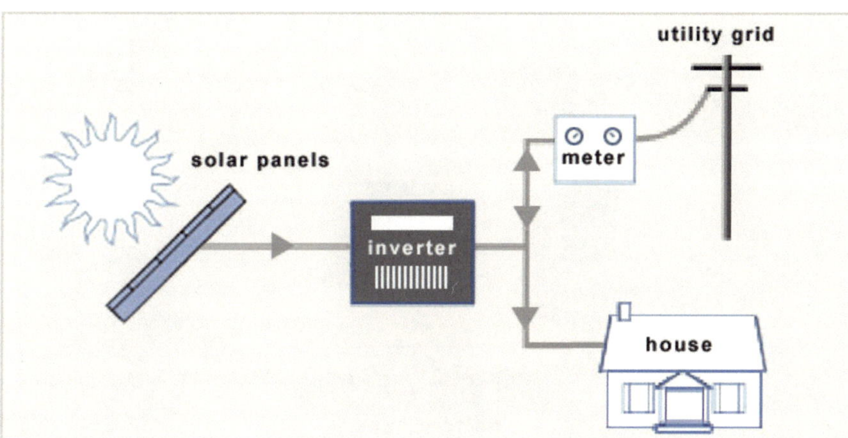

Figure 7.
Overview of the on-grid solar PV system.

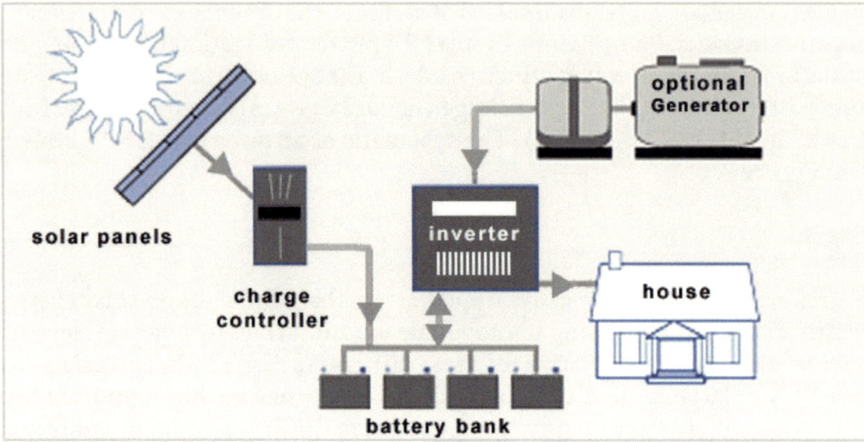

Figure 8.
Overview of the off-grid solar PV system.

control devices since PV is the only energy source, eliminating the need for interaction with other system components.

3.3 HYBRID system

The hybrid solar PV system combines the topologies of on-grid and off-grid systems. By leveraging the strengths and weaknesses of both systems, the hybrid topology provides a solution for supplying solar PV energy outside sunlight hours (e.g., to reduce load during peak times in the evening/night) while the PLN grid is still available. In this system, the battery storage capacity does not need to be as large as in off-grid systems. The battery storage capacity is determined by the amount of energy needed outside sunlight hours. This system also improves the reliability and quality of electricity supply from the existing grid (as a backup). The topology of a hybrid solar PV system is shown in **Figure 9**.

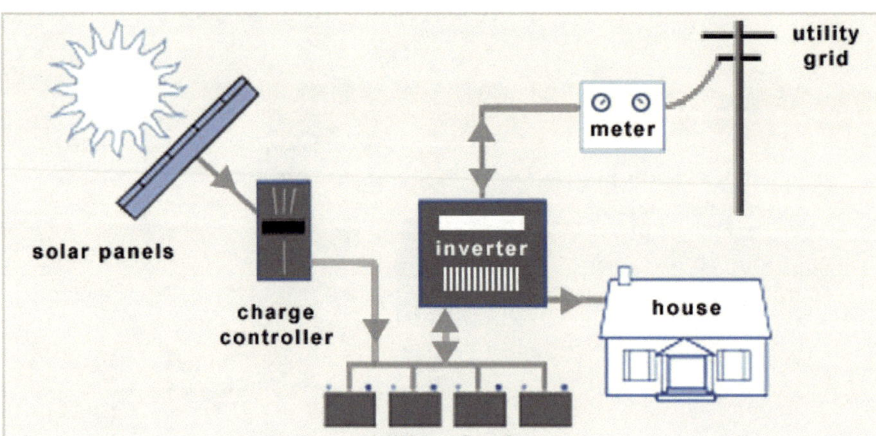

Figure 9.
Overview of the hybrid solar PV system.

For efficient operation, diesel generators must operate at 80–90% of their rated capacity. If a diesel generator operates below its capacity, it becomes inefficient. In most diesel powerplants systems operated in remote areas, peak load constitutes only a small percentage of the diesel generator's operating hours, even though the generator is designed to meet predicted peak loads.

To replace existing diesel powerplants, particularly in isolated networks, one candidate is the implementation of hybrid power generation technology combining diesel generators, photovoltaics, and batteries for remote and underdeveloped areas. These areas typically have low electricity demand and are far from primary energy sources. Various renewable energy sources are currently used to supply different applications in rural and urban areas. Hybrid systems using photovoltaics (PV) are considered the most popular among other renewable systems. The main advantages of this technology are low maintenance costs and low pollutant emissions. Hybrid solar PV-diesel-battery power generation technology (solar PV hybrid-diesel-battery) enables improved system efficiency, power reliability, and reduced energy storage requirements for standalone powerplants. Such hybrid systems have been widely implemented to reduce the consumption of increasingly expensive and difficult-to-obtain fossil fuels in remote and island areas.

By combining (hybridizing) with solar PV and batteries, diesel generator operations are expected to be more efficient, as daytime loads can be supplied by solar PV, with excess energy stored in batteries. During peak load conditions, typically at night, the diesel generator is operated, making it more efficient as it runs at around 80–90% of its rated capacity. The hybrid solar PV system is designed to meet several criteria, such as:

- Reliability: Ensuring continuous power supply.
- Cost-effective: Balancing initial costs with system operating costs.
- Efficiency: Maximizing overall system efficiency.
- Low maintenance: Minimizing system maintenance.

This study aims to recommend an optimal design model for a hybrid solar diesel powerplant using a battery bank to minimize costs and enhance system reliability. The general configuration and operation of a hybrid solar PV system are illustrated in the diagrams below (**Figures 9**, **10**, and **11**).

Key factors in the design calculations for a hybrid solar PV system include:

- Daily load curve
- Renewable energy source profile
- Dispatch strategy
- System power capacity planning.

Low load condition: The diesel generator does not operate, and the PV system and battery operate through an inverter that converts DC power to AC power (**Figure 10**).

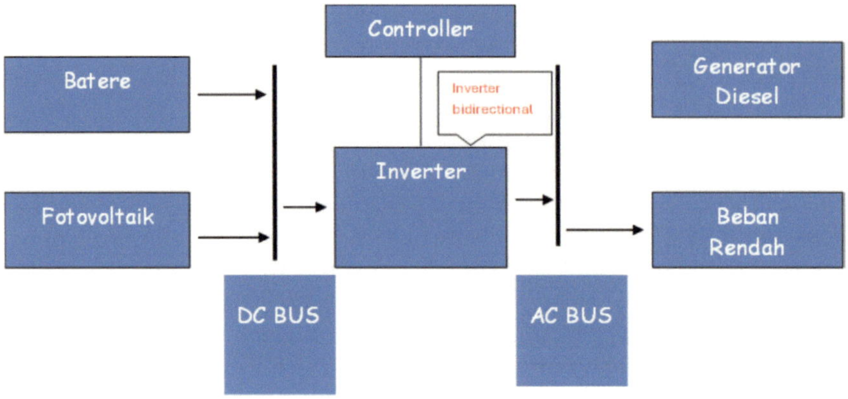

Figure 10.
Low load condition.

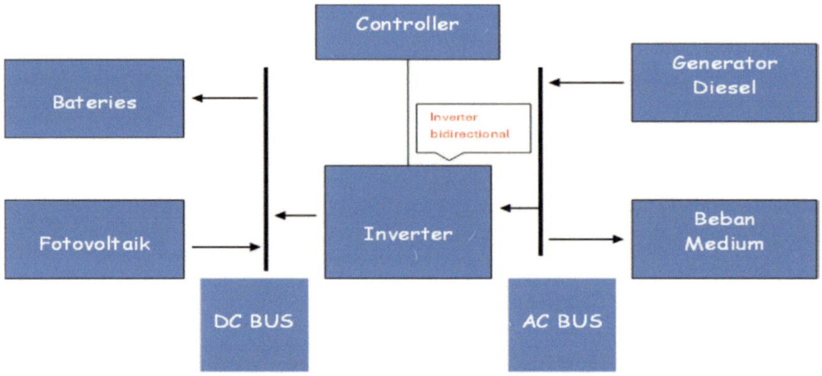

Figure 11.
Medium load condition.

Medium load condition: The diesel generator operates at optimal load to meet the load and simultaneously charge the battery through an inverter that converts AC power to DC power (**Figure 11**).

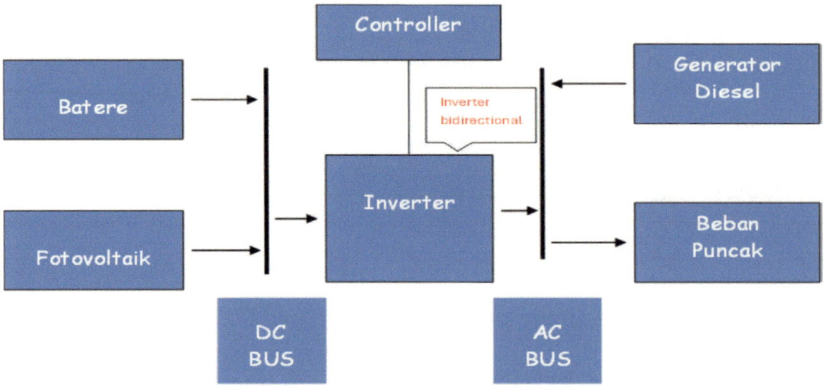

Figure 12.
Peak load condition.

DOI: http://dx.doi.org/10.5772/intechopen.1012089

Peak load condition: The diesel generator operates at optimal load to meet the load in parallel with an inverter that converts DC power from the battery and PV system to AC power (**Figure 12**).

4. Development of solar PV in Indonesia

This section focuses on the development of solar PV in Indonesia, highlighting the growing use of solar technology and advancements in local industry through increased local content. The development of solar powerplants (PLTS) in Indonesia is closely tied to its geographical location on the equator, which provides immense solar energy potential. Solar PV in Indonesia is distributed across various regions, with the three largest located in West Java, North Sulawesi, and East Nusa Tenggara. The Cirata Solar PV in West Java is the largest in Southeast Asia, with a capacity of 192 MWp. Additionally, there is the Likupang Solar PV in North Sulawesi with a capacity of 21 MWp and the Oelpuah Solar PV in East Nusa Tenggara with a capacity of 5 MWp. The following is an overview of the potential for solar PV development in Indonesia, focusing on geographical aspects, solar energy potential, completed projects, and development challenges:

Geographic advantage: Indonesia's equatorial location allows consistent sunlight exposure throughout the year, with an average irradiation duration of 10–12 hours per day. Solar radiation intensity in Indonesia ranges around 4.8 kWh/m^2/day, significantly higher than in subtropical or high-latitude regions.

Solar energy potential: According to the 2022 Indonesia Energy Outlook by the National Energy Council (DEN), Indonesia's solar energy potential reaches 3294 GW (3.294 TW), making it one of the largest renewable energy sources in the country. This potential is distributed across all regions, particularly in areas with vast land and minimal cloud cover, such as Java, eastern Sumatra, Kalimantan, Sulawesi, and eastern Indonesia.

Regional variations: Areas like East Nusa Tenggara (NTT), East Kalimantan, and Papua have higher potential due to lower rainfall and more consistent clear weather compared to western regions like western Sumatra or West Kalimantan. **Figure 13** illustrates the solar PV potential in Indonesia [28].

Figure 13.
Potential of solar PV in Indonesia [28].

Other notable projects include the 145 MWp Saguling Floating Solar PV (under development) and Bali's 2 × 25 MWp solar plants, reflecting diverse applications of solar PV across ground-mounted, floating, and rooftop systems.

Despite its potential, solar PV development in Indonesia faces several challenges: Intermittency issues require enhanced grid flexibility and battery energy storage systems (BESS). Frequent regulatory changes, such as revisions to local content requirements (TKDN), create investor uncertainty. While MEMR Regulation 11/2024 relaxed TKDN rules for foreign-funded projects, consistent policies are needed to sustain growth. Limited domestic production of key components like inverters and batteries results in low TKDN levels, increasing reliance on imports. Land acquisition for large-scale solar projects is complex due to competing land-use demands. Floating PLTS mitigates this but requires additional technical expertise.

4.1 Applied technology

From a technological perspective, several solar module technologies have been implemented or installed in Indonesia. Silicon-based solar modules are the most dominant technology in Indonesia, used in large-scale solar PV, rooftop solar PV, and off-grid systems in remote areas. Monocrystalline silicon modules, particularly with technologies like passivated emitter and rear cell (PERC), are widely used due to their high efficiency (22–24%) and market availability. The Cirata Solar PV (192 MWp), the largest floating solar PV in Southeast Asia, uses monocrystalline silicon modules. Likewise rooftop solar PV in residential, commercial, and industrial sectors also uses silicon modules. Off-grid projects in remote villages, such as Pertamina's Energy Independent Village program, also use silicon modules for solar home systems (SHS). The advantages of using these modules include competitive pricing, mature technology, and the availability of local components (such as aluminum frames and glass) that support the domestic component level (TKDN). However, challenges include local modules being less competitive than imports (local prices at 15–17 cents/kWh vs. imports at 10–12 cents/kWh), and the relaxation of TKDN from 40 to 20% has made imported products more dominant.

Bifacial solar module technology is starting to be implemented in Indonesia, particularly in large-scale projects like floating and ground-mounted solar PV. These modules absorb light from both sides, increasing energy yield by 10–20%. Examples include the Cirata floating Solar PV and the Saguling Solar PV (92 MWp). Bifacial technology maximizes energy production from direct sunlight and water reflections. Some commercial projects by PT Indonesia Solar Global (ISG) and other companies reportedly use bifacial modules. The advantages include suitability for high-reflectance environments like water or bright surfaces and increased energy efficiency. However, challenges include higher initial costs compared to monofacial modules and the need for optimized system design to utilize the module's rear side.

Thin-film technologies (such as CIGS or CdTe) and perovskite have not been widely implemented commercially in Indonesia. Thin-film is occasionally used for specific projects, such as building-integrated photovoltaics (BIPV), but on a very limited scale. Perovskite remains in the research stage and has not reached commercialization. No large-scale documented projects exist for thin-film or perovskite in Indonesia. Some companies, like Alva Energi, may be experimenting with these technologies for small-scale projects or research. The advantages of thin-film include flexibility and lightweight properties, suitable for BIPV applications, while perovskite offers potential efficiencies exceeding 30% in laboratory settings.

Figure 14.
Application of solar PV-BESS technology in Indonesia [29].

Battery technology developments have also been applied in solar PV construction in Jambi. This solar PV has a capacity of 643.8 kWp and operates off-grid, with a battery energy storage system (BESS) of 1 MWh packaged in a 20-foot container. Solar energy generated during the day is stored in the battery and released as needed. Based on technical analysis using the MCDM AHP TOPSIS method, the most optimal alternative is the lithium-ion battery (LIB), with the highest global weight value (2.875) due to its superiority in two priority criteria (**Figure 14**) [30].

4.2 Placement of solar modules

In terms of solar module placement, there are rooftop solar PV, ground-mounted solar PV, and floating solar PV. The adoption of rooftop solar PV has grown rapidly since the introduction of the net-metering scheme through Regulation of the Minister of ESDM No. 49/2018, although growth is still slower than the government's target (3.6 GWp by 2025). The installed capacity of rooftop solar PV reached 80.45 MWp by 2022, with 6522 users. Residential installations are prominent in Banten (2997 users, 12.4 MWp), Jakarta, and Bali, provided by companies like ATW Solar (over 3000 residential installations) and Inecosolar (nearly 0.5 MWp in Bali and East Java). Commercial and industrial sectors, such as those offered by Suryanesia with a solar-as-a-service model requiring no initial investment, are also growing. Advantages include flexibility for various roof types (tiles, concrete, and metal), electricity cost savings of 15–50%, and support for urban energy transitions. Current challenges include PLN's quota restrictions (10–15% capacity in 2022), the elimination of net-metering (exported surplus not credited in bills), and low public awareness.

Floating solar PV is a strategic innovation in Indonesia due to land constraints. Indonesia has significant potential for floating solar PV in reservoirs, with several projects already implemented. The Cirata Solar PV (145 MWp, completed in 2022, now expanded to 192 MWp) in Cirata Reservoir, West Java, is the largest pilot project in ASEAN. The Saguling Solar PV (92 MWp), funded by Proparco and partners, is targeted to reduce 63,100 tons of CO2/year. Plans for a floating solar PV in IKN (50 MWp) include exemptions from TKDN regulations to accelerate construction.

The concept of floating solar PV addresses land limitations, offers higher efficiency due to natural water cooling, and has significant potential (reservoirs and lakes in Indonesia support large-scale development). Challenges include lengthy import permit processes (up to 7 months) and the 60% TKDN requirement starting in 2025, which may hinder projects if the local industry is not ready.

4.3 Industry and TKDN of solar PV components in Indonesia

The calculation of TKDN for solar PV is divided into 3 (three) parts, namely the TKDN value for standalone distributed solar PV, standalone centralized, and centrally connected. The TKDN achievement for solar PV in 2022 and 2023 was 44.17% and 47.28%, respectively, increasing to 86.81% in 2024 [31]. Solar PV component equipment is specifically produced by the solar PV component industry, utilizing different technologies and industrial classifications. Components such as solar panels, controllers, inverters, batteries, and networks can now be produced in Indonesia. However, they generally still rely on imported raw materials due to various economic reasons.

4.3.1 Industry and TKDN of solar modules

Among the solar PV components with a long production chain and dependence on foreign industries are solar panels. Producing solar panels or modules requires a lengthy industrial process involving significant investment and industrial risks. To function as a solar PV, solar modules must be integrated with other components in a single system. Based on data from the Ministry of Industry (Kemenperin) as of March 2025, there are approximately 33 companies producing solar modules, with a total annual production capacity of 4.3 gigawatt peak (GWp). However, most of these companies are still at the module assembly stage and have not developed a full upstream-to-downstream industry, such as solar cell and silicon wafer production.

Of these 33 companies, 22 have TKDN certificates, and 14 are members of the Indonesian Solar Module Manufacturers Association (APAMSI*), as listed in **Table 1**.

The production capacity of most APAMSI members exceeds 40 MWp per year. The majority of APAMSI member industries are located in western Java (West Java, Jakarta, and Banten provinces). This is due to several factors, including more advanced industrial infrastructure, good access to resources, supply, and markets, and the availability of skilled labor.

4.3.1.1 TKDN of the solar module industry

From **Table 1**, it is evident that 22 solar module companies have obtained TKDN certificates from the Ministry of Industry. The produced solar modules range from 10 to 670 Wp per module, with monocrystalline and polycrystalline types. The majority of solar module manufacturers in Indonesia have TKDN certificates above 40%. According to Government Regulation No. 29/2018 on industrial empowerment, products with a TKDN value above 25% receive a 25% price preference in competition with foreign products. Products with a TKDN of 25% and a company benefit weight (BMP) of 15% are mandatory for use in government-funded projects. With the above data, domestic solar module products have significant potential for use in solar PV development in Indonesia.

No.	Company name	Specification	Crystalline type	TKDN value	Capacity (MWp/Year)
1.	PT. Abhinaya Rawi Sahwahita	100–540 Wp	Mono	38.00%	
		10–200 Wp	Poly	38.00%	
2.	PT. Adyawinsa Electrical & Power*	410–660 Wp	Mono	46.50%	40
		155–405 Wp	Mono	47.50%	
		20–330 Wp	Poly	47.50%	
3.	PT. Adzla Gemilang Sejahtera			10.00%	
4.	PT. Apollo Solar Indonesia	550–670 Wp	Mono	40.00%	
5.	PT. Avecode Internasional	60–540 Wp	Mono	47.50%	
6.	PT. Azet Surya Lestari*	400–550 Wp	Mono	40.00%	45
		20–300 Wp	Mono	47.50%	
7.	PT. Bernadi Utama*	550 Wp	Mono	40.00%	30
		510 Wp	Mono	40.00%	
8.	PT. Deltamas Solusindo*	30–550 Wp	Poly	47.50%	30
		30–550 Wp	Mono	47.50%	
9.	Hit Internasional	550 Wp	Mono	33.50%	
10.	PT. Indodaya Cipta Lestari*	100–550 Wp	Mono	45.50%	30
11.	PT. Jembo Energindo*	95–550 Wp	Mono	54.48%	50
		150–330 Wp	Poly	54.48%	
12.	PT. LEN Industri (Persero)*	210–400 Wp	Mono	46.62%	75
13.	PT. Lesso New Energy	585–670 Wp	Mono	39.55%	
		390–555 Wp	Mono	38.00%	
14.	PT. Pana Indo Alkestama	250–455 Wp	Mono	40.00%	
15.	PT. Sankeindo*	270–330 Wp	Poly	46.50%	45
		185–590 Wp	Mono	46.50%	
16.	PT. Santinilestari Energi Indonesia*	10–400 Wp	Mono	47.50%	50
17.	PT. Sinar Energi Gemilang	50, 100, 150, 200, 300 Wp		30.00%	
		50, 100, 150, 200, 300 Wp		38.00%	
18.	PT. Surya Utama Putra*	30–550 Wp	Mono	47.50%	45
		30–320 Wp	Poly	47.50%	
19.	PT. Swadaya Prima Utama*	200–560 Wp	Mono	47.50%	40
		100–360 Wp	Poly	47.50%	
20.	PT. Techlan Solar Indonesia	30–455 Wp	Mono	47.50%	
		30–350 Wp	Poly	47.50%	
21.	PT. Waskita Adhi Sejahtera		Mono	40.00%	
22.	PT. Wijaya Karya Industri Energi*	340 Wp	Mono	38.50%	50

No.	Company name	Specification	Crystalline type	TKDN value	Capacity (MWp/Year)
23.	PT. Cahaya Mas Cemerlang*				10
24.	PT. Sky Energy Indonesia Tbk.*				200

Source: P3DN Kemenperin 2025, link http://tkdn.kemenperin.go.id [32].

Table 1.
Solar module industry in Indonesia.

4.3.2 Industry and TKDN of inverters

Inverter production in Indonesia is almost nonexistent, as most inverters used are imported. The only local industry producing inverters is the state-owned PT. Len Industri in Bandung. Its products have a limited capacity of up to 15 kW for both off-grid and on-grid inverters, as shown in **Figure 15**. The TKDN achievement value for inverter products is currently unavailable.

4.3.3 Industry and TKDN of batteries

To support clean energy provision, the industry plays a crucial role, as mass production of secondary batteries occurs in the industrial sector. In Indonesia, industries engaged in battery production are limited and generally produce secondary batteries of the starting, lighting, and ignition (SLI) type for vehicles. As the market for electric vehicles and alternative energy powerplants develops in the future, the demand for secondary batteries with high capacity and energy density will be essential. Consequently, the battery industry in Indonesia will also shift toward producing secondary batteries with these characteristics. The TKDN achievement value for battery products is currently unavailable. Below is the data on battery production industries in Indonesia (**Table 2**).

Of the above industries, only PT. Nippress Tbk is a domestically owned company, while the others are generally foreign-invested (PMA), making their energy storage industry development policies dependent on their parent companies. Currently, PT. Nippress Tbk has successfully produced lithium batteries for electric vehicles and is developing lithium batteries for mobile BTS. However, in the production and development of these lithium batteries, PT. Nippress is limited to assembly, as lithium cells are still imported from abroad. This indicates that Indonesia is structurally not yet

Battery Inverter 5kW
Off Grid Inverter Bidirectional 15 kW
On Grid Inverter 5 kW

Figure 15.
Inverter produced by PT Len Industri Bandung [22].

No.	Company name	Product name	Description
1.	PT. Indobatt	NGS, NEO, Volcano	Factory located in Krian, East Java. Produces batteries for cars and motorcycles.
2.	PT. Gramitama Battery	GBI, GS Premium, OSAKA, Yama	Factory located in Sidoarjo, East Java. Produces batteries for cars and motorcycles.
3.	PT. Yuasa Battery	Yuasa Pafecta, Yuasa Maintenance Free, Yuasa Hybrid, Yuasa Yumicron, Yuasa VRLA, Yuasa Super MF	Factory located in Tangerang, Banten. Produces batteries for cars and motorcycles.
4.	PT. International Chemical Industry	ABC Alkaline, ABC Super Power, ABC New Special, ABC Dry Cell, ABC Economy	Factory located in Cengkareng, Banten. Produces dry cell batteries for low-voltage applications.
5.	PT. World Star Battery Indonesia	WS Worldstar	Factory located in Surabaya, East Java. Produces batteries for cars and motorcycles.
6.	PT. Leoch Battery Indonesia	LP, LPX, LHR, LPL, LPF, DJW, LPG, LPG-FT, LPC	Factory located in Daan Mogot, Tangerang, Banten. Battery applications for electric vehicles, motorcycles, cars, UPS, and others. Also produces chargers, power inverters, and golf carts.
7.	PT. Nippress, Tbk	NS	Factory located in Bogor, West Java. Battery applications for cars, motorcycles, golf carts, and industrial batteries.

Table 2.
Data on battery manufacturing industries in Indonesia [22].

independent in developing energy storage technology. However, given the sufficient reserves of raw materials to support the lithium battery industry, as discussed in the previous chapter, Indonesia has the potential to build a lithium battery industry with domestically sourced raw materials.

5. Risk mitigation

This section focuses on risk mitigation strategies for solar PV, addressing the challenges associated with solar energy's intermittency and ensuring reliable power supply. It emphasizes the importance of comprehensive risk management in ensuring the stability and sustainability of solar power generation, particularly through hybrid systems and proactive planning.

One of the main challenges of solar PV is its dependence on unpredictable weather conditions, particularly at night or during cloudy weather. Solar energy cannot be utilized continuously, and when sunlight intensity is low, solar PV cannot generate electricity optimally. Therefore, to maintain energy supply stability, diesel powerplants remain necessary as a backup or additional power source that operates independently of weather conditions. A viable solution is to hybridize diesel powerplants and solar PV. This hybrid system combines the strengths of both technologies, where solar PV generates electricity during the day, reducing reliance on fossil fuels, while diesel powerplant is available to meet electricity needs when solar PV is not operational.

Risk mitigation in hybrid solar PV-diesel powerplants systems is critical to ensure the reliability and stability of the energy supply, especially given solar PV's dependence on unpredictable weather conditions. Below are risk mitigation methods with a maximum hybrid solar PV power supply portion of 30% of the power generated by diesel powerplant.

5.1 Identification of risk causes

In a hybrid power generation system combining solar PV and diesel powerplant, various factors can pose risks to performance and operational sustainability. Identifying these risk causes is essential to ensure the overall system's stability and efficiency. Key risk causes in hybrid solar PV projects include the quality of solar PV electricity (technology), smart grid device performance, physical conditions and project location, and external factors such as weather conditions and energy sustainability (fuel supply). Additionally, challenges in initial investment costs and limited community empowerment opportunities (economic impact) are significant risk factors, particularly in the context of a community-based ecosystem scheme. Understanding these risk causes allows for the design of mitigation measures to reduce potential negative impacts, ensuring the hybrid power generation system operates more reliably and sustainably.

5.2 Risk assessment

Risks are assessed based on the multiplication of severity (consequence) and the likelihood of occurrence (numerical ranking) for each identified risk. Below are the risk components developed into five main categories with their subcomponents:

1. *Fuel supply assurance* includes quantity assurance, quality assurance, and timely supply assurance.

2. *Solar PV technology* includes technical feasibility, technology availability, reliability, productivity, solar PV technology operations, and maintenance and spare parts procurement.

3. *Investment* includes economic feasibility, investment value per plant (300 kWp), local government support, PLN investment interest, and IPP investment interest.

4. *Economic impact* includes economic impact on PLN, economic impact on IPP, community economic improvement, and regional economic improvement.

5. *Environment* includes pollutant products, environmental impact throughout the product or service lifecycle, regional economic driver ecosystem, and community economic driver ecosystem.

5.3 Risk mitigation

Risk mitigation involves efforts to reduce or eliminate risks through preventive measures to ensure identified risks do not hinder goal achievement. Not all risks require mitigation. Low risks can be tolerated without affecting goal achievement (risk appetite). Below are mitigation measures for medium and high risks. **Table 3** below presents the risk mitigation analysis, including risk description, risk cause, risk impact, and risk mitigation.

DOI: http://dx.doi.org/10.5772/intechopen.1012089

No.	Risk description	Risk cause	Risk impact	Risk mitigation
1	Medium risk			
	Fuel supply assurance (solar radiation) includes quantity, quality, and timely supply.	Sumbawa's rainy season, with 5 months of rain, affects solar radiation intensity.	Reduced PV electricity supply during the rainy season increases diesel powerplant operation.	Utilize accurate rainy season data for planning diesel powerplant operation patterns as a backup for reduced PV power during the rainy season.
	Solar PV technology includes technical feasibility, reliability, and productivity.	Inverter, battery, and smart grid device malfunctions.	Reduced reliability, increasing diesel powerplant load.	Select high-quality, tested, and certified equipment. Perform O&M according to manufacturer SOPs.
	Investment includes PLN interest.	High investment costs.	Long investment payback period.	Link solar PV operation to mandatory EBT usage value.
	Environment includes regional and community economic driver ecosystems.	Solar PV operation provides limited business opportunities for community fuel supply.	Community ecosystem goals are not directly achieved through solar PV operation.	Conduct socialization and education to the community about the benefits of solar PV for regional electricity supply.
2.	High risk			
	Solar PV technology: Operations, maintenance, and spare parts procurement during issues.	Spare parts are not available locally, especially in remote areas.	Increased diesel powerplant operating time.	Provide backup spare parts for critical equipment.
	Investment: Economic feasibility and IPP interest.	Low estimated economic feasibility.	Reduced investment appeal.	Calculate economic feasibility with various incentives.
	Economic impact: For PLN, IPP, and community economic improvement.	Low estimated economic feasibility, unattractive profits, and limited community participation opportunities.	Unattractive to investors, community ecosystem goals not achieved.	Provide special incentives for investors and conduct community socialization.

Table 3.
Solar PV risk mitigation analysis.

Based on the above analysis, the risk mitigation measures are as follows:

1. Fuel supply assurance: Utilize accurate rainy season data to plan diesel powerplant operation patterns as a backup for PV energy during the rainy season.

2. Solar PV technology: Select high-quality, tested, and certified equipment. Perform O&M according to manufacturer SOPs.

3. Investment: Link solar PV operation to mandatory EBT usage to accelerate investment payback.

4. Environment: Conduct socialization and education to the community about the benefits of solar PV for electricity supply.

5. Spare parts and maintenance: Provide backup spare parts for critical equipment and calculate economic feasibility with various incentives to attract investment.

6. Conclusion

Indonesia has set an ambitious target to achieve a significant share of its energy mix from new and renewable energy (EBT), with solar PVs (PLTS) playing a key role. Based on the findings of this research, the target appears achievable due to the following factors:

- *Geographic advantage*: Indonesia's equatorial location offers abundant solar energy potential, with an average irradiation of 4.8 $kWh/m^2/day$, ensuring consistent sunlight exposure crucial for large-scale solar energy generation.

- *Technological advancements*: Continuous improvements in solar technology, including increased solar module efficiency, the adoption of bifacial modules, and reduced costs for solar cells, modules, and energy storage solutions, have made solar power increasingly cost-effective and scalable.

- *Increasing installed capacity*: By 2023, Indonesia's installed PLTS capacity reached 573.8 MW, with major projects such as the Cirata Floating PLTS (192 MWp). This progress demonstrates the country's potential for growth, and with Indonesia's solar energy potential estimated at 3294 GW, there is ample opportunity to expand capacity further to meet the energy mix target.

- *Local industry development (TKDN)*: Indonesia's success in increasing the TKDN for solar components (44.17% in 2022, 47.28% in 2023, and 86.81% in 2024) supports the growth of a domestic solar industry, reducing reliance on imports. This strengthens the local energy sector and accelerates solar power deployment.

- *Government support*: The Indonesian government's policies, such as the Net-Metering Scheme for rooftop PLTS, infrastructure development for renewable energy, and clear policies for expanding solar energy capacity, align with the goal of achieving a significant share of the energy mix from renewable sources. This regulatory framework will further facilitate the growth of solar energy.

PLTS offers a compelling opportunity for Indonesia's future energy needs. The combination of the country's advantageous geographic position, advancements in solar technology, increasing installed capacity, and strong government support positions solar power as a sustainable and cost-effective solution to meet the growing demand for energy.

Conflict of interest

The authors declare no conflicts of interest.

Author details

Agus Krisnowo[1,2]*, Oo Abdul Rosyid[3], Iwan Sudrajat[2], Jaizuluddin Mahmud[2], Gani Soehadi[2], Ignatius Subagjo[2], Irawan Santoso[2], Mohamad Dahsyat[2], Yustiar Gunawan[2], Syaeful Karim[4], I Made Astawa[4], Ugay Sugarmansyah[5] and Susalit Setya Wibowo[2]

1 Ph.D Student School Interdisciplinary Management – Sepuluh Nopember Institute of Technology Surabaya, Indonesia

2 Research Center for Process and Manufacturing Industry Technology-BRIN, Jakarta, Indonesia

3 Research Center for Energy Conversion and Conservation-BRIN, Jakarta Indonesia

4 Research Center for Electronica-BRIN, Jakarta, Indonesia

5 Research Center for Sustainable Production System and Life Cycle Assessment-BRIN, Jakarta Indonesia

*Address all correspondence to: aguskrisnowo@yahoo.com

References

[1] Government Regulation Number 79/2014 Concerning National Energy Policy (KEN). 2014

[2] Presidential Regulation Number 22/2017 Concerning the National Energy General Plan (RUEN). 2017

[3] Presidential Regulation Number 98/2021 Concerning the Implementation of Carbon Economic Values for the Achievement of Nationally Determined Contribution Targets and Control of Greenhouse Gas (GHG) Emissions in National Development. 2021

[4] Ministry of Energy and Mineral Resources. Sun for Solar PV in Indonesia. 2012. Available from: https://www.esdm.go.id/id/media-center/arsip-berita/matahari-untuk-plts-di-indonesia

[5] Ministry of Energy and Mineral Resources. Sun for Solar PV in Indonesia. 2012. Available from: https://www.esdm.go.id/id/media-center/arsip-berita/menteri-esdm-ungkap-strategi-penuhi-target-bauran-energi-dari-ebt

[6] Uly YA, Djumena E. High Demand, Ministry of ESDM Considers Increasing Rooftop Solar PV Quota. Kompas.com; 2025. Available from: https://money.kompas.com/read/2025/04/30/074100426/permintaan-tinggi-kementerian-esdm-kaji-tambah-kuota-plts-atap-?page=all

[7] Regulation of the Minister of Energy and Mineral Resources Number 2 of 2024 Concerning Rooftop Solar PV Connected to the Electricity Network for Holders of Electricity Supply Business Licenses for Public Interest. 2024

[8] Lincot D. The new paradigm of photovoltaics: From powering satellites to powering humanity. Comptes Rendus Physique. 2017;**18**(7-8):381-390. DOI: 10.1016/j.crhy.2017.09.003

[9] Widiarto D. Solar Cell Panel Technology as a Power Generator. 2018. Available from: https://www.diditwidiarto.com/2018/09/teknologi-panel-solar-cell-sebagai.html#google_vignette

[10] GSES. All about High Efficiency Solar Cells: PERC and Half Cut. 2020. Available from: https://www.gses.com.au/high-efficiency-cells-perc-and-half-cut/

[11] De Burca J. Perovskite Silicon Tandem Photovoltaics-Solar Energy. 2024. Available from: https://constructive-voices.com/perovskite-silicon-tandem-photovoltaics-solar-energy/

[12] Bonilla S. New Tandem Solar Cell Technology Breaks World Record in Energy Efficiency. 2024. Available from: https://theconversation.com/teknologi-baru-sel-surya-tandem-pecahkan-rekor-dunia-dalam-efisiensi-energi-242486

[13] Longi. 34.85% LONGi Breaks World Record for Crystalline Silicon-Perovskite Tandem Solar Cell Efficiency Again. 2025. Available from: https://www.longi.com/en/news/silicon-perovskite-tandem-solar-cells-new-world-efficiency/

[14] Shah J. Explained: Bifacial Solar Panels. 2022. Available from: https://www.saurenergy.com/solar-energy-blog/explained-bifacial-solar-panel-and-everything-you-need-to-know-about-it

[15] DS New Energy. CIGS Solar Cell Technology. 2019. Available from:

https://www.dsneg.com/info/cigs-solar-cell-technology-32692187.html

[16] del Barrio Castro L, Martorel JA, Opitz M, Sampedro M, Estrada E. Hybrid Renewable Ecosystems, Leveraging Synergies for Green Data Center Growth through Renewables. 2025. Available from: https://www.adlittle.com/en/insights/viewpoints/hybrid-renewable-ecosystems

[17] Al Haqq MR, Cholissodin I, Soebroto AA. Maximum power point tracking (MPPT) on solar panels under partial shading conditions with particle swarm optimization (PSO). Jurnal Pengembangan Teknologi Informasi dan Ilmu Komputer. 2021;**5**(8):3524-3537. Available from: http://j-ptiik.ub.ac.id

[18] Kristiadi D. South Korea Develops Solar Panel Recycling Technique. 2021. Available from: https://m.icasolar.com/support/blog/daur-ulang

[19] Maghfiroh OI. Recycling Unused Solar Panels, Efforts for Sustainable Utilization. 2023. Available from: https://zonaebt.com/daur-ulang-panel-surya-yang-tak-terpakai-upaya-pemanfaatan-dari-praktik-berkelanjutan

[20] Ministry of Industry and Information Technology. Increase Green and Efficient Resource Utilization, and Promote Recycling of Used Solar Moduls. 2025. Available from: https://www.metal.com/id/newscontent/103400159

[21] Wibowo AH. Historical Decline in Global Lithium Battery Costs Based on BNEF Survey. 2022. Available from: https://data.goodstats.id/statistic/penurunan-historis-biaya-baterai-lithium-global-berdasarkan-survei-bnef-F22GU

[22] Krisnowo A. Baseline TKDN Achievement & Implementation of Increased Use of Domestic Products in Power Generation, Transmission, Substations, and Distribution at PT. Pembangkit Listrik Negara, Technical Document. 2021

[23] Hassan et al. Design and implementation of single phase pure sine wave inverter using multivibrator IC. In: 17th UKSIMM-AMSS International Conference on Modelling and Simulation. Cambridge, UK: IEEE Xplore; 2015. Available from: https://uksim.info/uksim2015/data/8713a451.pdf

[24] Chang E-C. High-performance pure sine wave inverter with robust intelligent sliding mode maximum power point tracking for photovoltaic applications. Micromachines. 2020;**11**(6):585. DOI: 10.3390/mi11060585

[25] Yee atonergi. Understanding Grid-Tied Inverters: Working Principles and Benefits. 2024. Available from: https://atonergi.com/mengenal-inverter-grid-tied-prinsip-kerja-dan-keuntungannya/

[26] Chaudhari C et al. PV Performance Modeling Collaborative (PVPMC), CEC Inverter Test Protocol. Sandia National Laboratories; 2025. Available from: https://pvpmc.sandia.gov/modeling-guide/dc-to-ac-conversion/cec-inverter-test-protocol

[27] Rahman Y et al. Performance of grid tie inverter with load variations on PV-on grid module trainer. Jurnal Teknik Energi Elektrik, Teknik Telekomunikasi, & Teknik Elektronika. 2022;**10**(2):287-296. DOI: 10.26760/elkomika.v10i2.287

[28] Agarwal V et al. How to Power Indonesia's Solar PV Growth Opportunities. 2023. Available from: https://www.mckinsey.com/id/our-insights/how-to-power-indonesias-solar-pv-growth-opportunities

[29] Sun Energy. Solar PV Innovation with Microgrid Battery Energy Storage System in Indonesia's First and Largest Container. 2025. Available from: https://sunenergy.id/inovasi-plts-dengan-sistem-penyimpanan-energi-baterai-microgrid-dalam-kontainer-pertama-dan-terbesar-di-indonesia-id

[30] Susanto FD. Technical analysis and feasibility of implementing battery energy storage system for load shifting in the southern Sumatra PLN subsystem [Masters thesis]. Surabaya: Institut Teknologi Sepuluh Nopember; 2025

[31] Directorate General of New, Renewable Energy, and Energy Conservation. Ministry of Energy and Mineral Resources, Performance Report of the Directorate General of EBTKE. 2024

[32] Center for the Promotion of Domestic Product Use (P3DN). Ministry of Industry, Inventory List of Domestic Product Goods and Services; 2025. Available from: http://tkdn.kemenperin.go.id/

Chapter 5

Future Trends in DC-DC Conversion for Renewable Energy Systems

Mohammed Gronfula

Abstract

The purpose of this chapter is to investigate emergent trends, novel designs, and potential future pathways in DC-DC converter technologies as applied to renewable energy systems. It will offer a prospective examination of advanced converter configurations, control mechanisms, integration approaches, and application contexts anticipated to shape the next iteration of power electronics. The overarching aim is to provide readers with foundational knowledge of how technological advancements can optimize the performance metrics of renewable energy conversion systems, including efficiency, reliability, scalability, and intelligent operation, in response to increasing energy consumption and sustainability imperatives.

Keywords: advanced DC-DC converters, renewable energy integration, power conversion efficiency, wide-bandgap semiconductors, modular and hybrid topologies

1. Introduction

1.1 Overview of DC-DC conversion in renewable energy

As the global energy landscape shifts toward sustainability and decarbonization, renewable energy systems such as photovoltaic (PV) arrays, wind turbines, and fuel cells have emerged as vital sources of electricity. These sources inherently produce variable and often unregulated DC or AC voltages, which are rarely suitable for direct usage or integration into modern energy systems without intermediate power conditioning. In this context, DC-DC converters play a pivotal role in enabling the efficient interface between renewable energy sources and loads or energy storage systems by adjusting voltage levels, improving power quality, and providing isolation where necessary.

DC-DC converters are power electronic devices that convert a source of direct current (DC) from one voltage level to another. Their usage in renewable energy systems is crucial not only for voltage matching but also for implementing advanced control strategies such as maximum power point tracking (MPPT), energy storage management, and grid synchronization. For instance, in a PV system, the solar panel output varies depending on irradiance and temperature, and a DC-DC converter dynamically adjusts the operating point to extract the maximum power possible [1].

1.2 Importance of efficient power conversion

Efficient power conversion is fundamental to the performance and viability of renewable energy systems. As these systems scale up, even minor losses in conversion stages can result in substantial energy and economic inefficiencies. Modern DC-DC converters must deliver high efficiency across a wide operating range, maintain a compact size, and offer high power density. With the advent of wide-bandgap semiconductors such as silicon carbide (SiC) and gallium nitride (GaN), converter efficiency has improved significantly, achieving levels beyond 98% in optimized systems [2].

In addition, DC-DC converters are essential for integrating various subsystems in hybrid renewable energy architectures, which often involve multiple energy sources and storage elements. Their role is not limited to energy conversion; they also support advanced functionalities such as bidirectional power flow, isolation, fault tolerance, and protection. These capabilities are particularly important in applications like electric vehicle (EV) charging infrastructure, microgrids, and remote off-grid installations [3].

Traditional DC-DC converters, while effective for basic voltage regulation, are increasingly insufficient in meeting the evolving needs of decentralized, multisource renewable systems. There is a growing need for converters that support dynamic energy management, operate efficiently across a wide range of loads, and integrate with intelligent control and communication infrastructure. This chapter addresses these challenges by exploring advanced converter topologies and emerging technologies that align with current and future system requirements.

1.3 Scope and organization of the chapter

This chapter provides a comprehensive exploration of DC-DC converters, emphasizing their topological evolution, control strategies, and integration into renewable energy systems. It begins by presenting the *fundamentals of DC-DC converter operation*, highlighting classical and advanced topologies such as buck, boost, buck-boost, SEPIC, and interleaved converters. Special attention is given to isolated topologies like flyback and forward converters, which are essential for systems requiring galvanic isolation.

Subsequently, the chapter delves into *modern control techniques* used in DC-DC converters, ranging from traditional linear controllers to digital and AI-assisted control schemes. These control strategies are critical for achieving fast dynamic response, robustness under parameter variation, and optimized energy extraction.

The next section focuses on the *integration of DC-DC converters in renewable energy systems*, covering practical deployment in PV, wind, fuel cell, and hybrid systems. Integration challenges, including system-level optimization, component interactions, and reliability under dynamic conditions, are discussed with practical examples.

Finally, the chapter presents *emerging trends and future directions*, including smart converters, high-frequency operation, reconfigurable topologies, and their role in next-generation DC microgrids and smart energy systems.

By covering both theoretical and application-oriented aspects, this chapter aims to equip researchers, engineers, and students with a solid understanding of the evolving landscape of DC-DC conversion in renewable energy technologies.

The primary objective of this chapter is to provide a forward-looking examination of DC-DC converter technologies in renewable energy systems by highlighting advanced topologies, intelligent control strategies, integration frameworks, and future trends. This chapter aims to equip researchers and practitioners with insights into how emerging designs and techniques can address evolving performance, efficiency, and scalability demands in sustainable power systems.

2. Fundamentals, classification, and performance metrics of DC-DC converters in renewable energy systems

2.1 Basic principles of DC-DC conversion

DC-DC converters are critical power electronic components utilized to transform one direct current voltage level into another. Their functioning relies on switching technologies—commonly employing transistors like Metal–Oxide–Semiconductor Field-Effect Transistors (MOSFETFs) or Insulated-Gate Bipolar Transistors (IGBTs)—to cyclically store and release energy through inductors, capacitors, and transformers. This switch-mode operation facilitates highly efficient conversion across diverse voltage and load conditions, an essential requirement for renewable energy systems.

The fundamental principle underlying DC-DC converters involves modulating the duty cycle, which represents the ratio of the switch-on time to the total switching period, to regulate the output voltage. By rapidly activating and deactivating the switch, the converter modulates the energy transfer in discrete packets, which are then smoothed using output filters. In contrast to linear regulators that dissipate excess energy as heat, switch-mode DC-DC converters can achieve efficiency levels exceeding 90% by minimizing energy losses [4].

This switching mechanism introduces inherent trade-offs, such as switching losses, electromagnetic interference (EMI), and output voltage ripple, which necessitate careful design and control strategies to manage effectively (see **Figure 1**).

2.2 Classification of DC-DC converters

DC-DC converters can be categorized into two main topological classifications: non-isolated and isolated, determined by the presence or absence of electrical isolation between the input and output.

2.3 Non-isolated converters

Non-isolated converters are frequently employed in low- to medium-power applications, such as those found in photovoltaic modules, battery-powered systems, and electric vehicles. The primary topologies within this category include:

- *Buck converter:* Steps down voltage.
- *Boost converter:* Steps up voltage.
- *Buck-boost converter:* Provides inverted output voltage with step-up or step-down capability.
- *SEPIC and Cuk converters:* Offer non-inverting or inverting outputs with improved ripple characteristics.

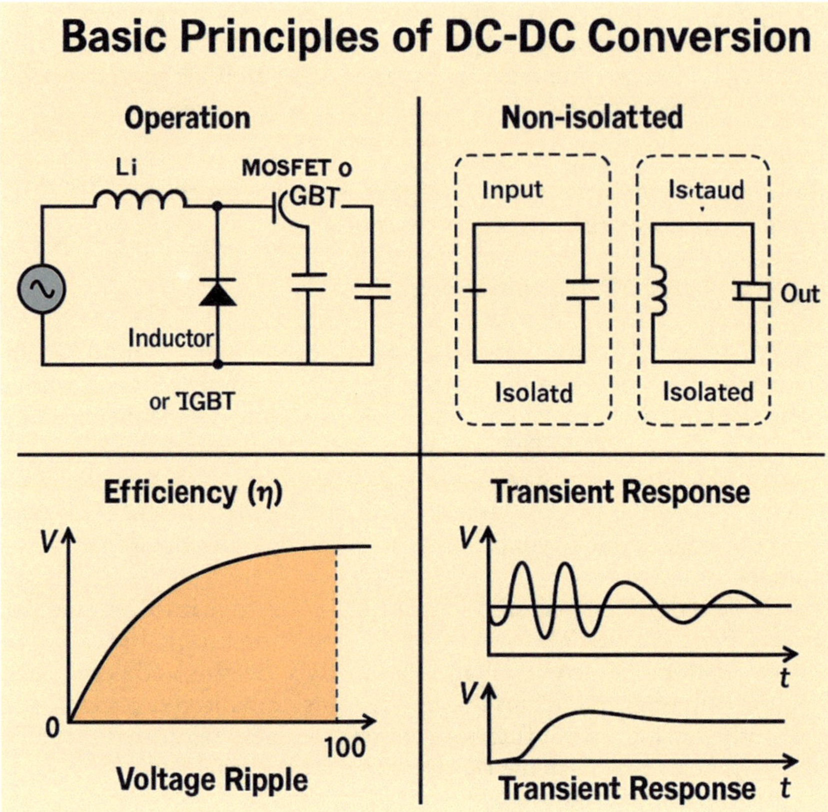

Figure 1.
Basic principles and key metrics of DC-DC converters.

Non-isolated DC-DC converter topologies tend to be less complex, more space-efficient, and more energy-efficient at lower power levels, rendering them well-suited for applications that do not necessitate electrical isolation between the input and output [5].

2.4 Isolated converters

Isolated converters utilize a transformer to provide galvanic isolation and enable voltage scaling across broad ranges. Some widely adopted topologies in this category include:

- *Flyback converter:* Simple and widely used for low-power isolated applications.
- *Forward converter:* Offers better efficiency at higher power levels.
- *Push-pull, half-bridge, and full-bridge converters:* Suitable for midium to high power applications with symmetrical operation and better transformer utilization.

Isolated converters are crucial in grid-connected systems and safety-critical applications where preventing ground loop currents or ensuring operator safety is paramount [6].

Additionally, *multi-port converters* and *modular topologies* are gaining prominence in renewable systems to enable integration of multiple sources and storage units with high flexibility and scalability (see **Figure 1**).

2.5 Key performance metrics

Design and selection of a DC-DC converter depend on several performance metrics that influence system efficiency, stability, and dynamic response.

2.5.1 Efficiency (η)

Efficiency (η) is defined as the ratio of output power (P_out) to input power (P_in), expressed as a percentage:

$$\text{ç} = (P\_out / P\_in) \times 100\% \qquad (1)$$

High efficiency reduces energy losses and thermal management requirements. In renewable systems, where power generation is often intermittent and limited, maintaining high conversion efficiency is essential to optimize energy utilization [7].

2.5.2 Voltage ripple

Voltage ripple refers to the AC variation superimposed on the DC output voltage. Excessive ripple can lead to undesirable effects on sensitive loads, reduced battery life, or increased electromagnetic interference. The ripple is influenced by factors such as switching frequency, filter design, and load characteristics. Effective filtering and layout techniques are necessary to minimize ripple and meet regulatory standards.

2.5.3 Transient response

Transient response is the converter's ability to respond to sudden changes in load or input voltage without deviating excessively from the desired output. Fast and well-damped transient behavior is critical in renewable systems, especially in grid-tied applications or fluctuating sources like wind and solar power. It ensures stability and prevents system shutdown or power quality issues.

Other considerations include *power density*, *thermal performance*, *EMI compliance*, and *control loop stability*, all of which must be optimized depending on the application.

3. Converter topologies for renewable energy systems

The choice of an appropriate converter topology is essential for enhancing the performance, efficiency, and dependability of renewable energy systems. Converter topologies should be selected based on the characteristics of the power source, load requirements, isolation needs, and control objectives. This section provides a detailed examination of the most pertinent converter types, which can be categorized into non-isolated, isolated, multi-port, and switched passive component converters (see **Figure 2**).

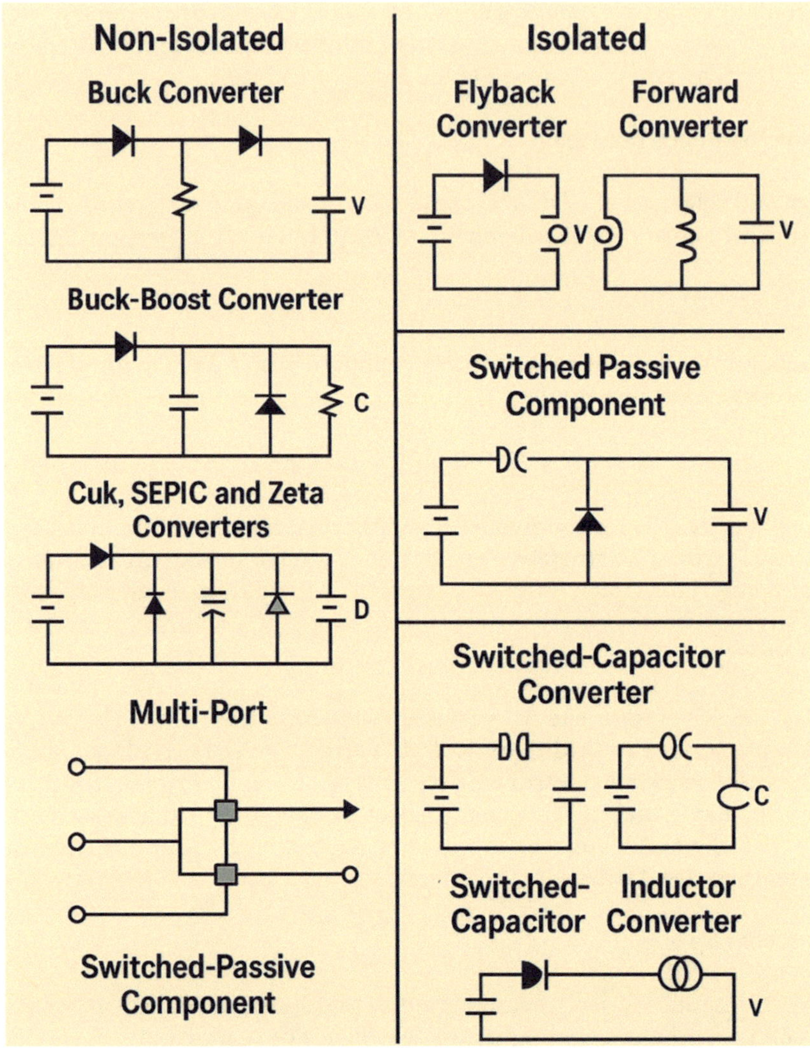

Figure 2.
Converter topologies for renewable energy systems.

3.1 Non-isolated converters

3.1.1 Buck converter

The buck converter is a step-down DC-DC converter utilized when the output voltage needs to be lower than the input voltage. It functions by rapidly switching a transistor to regulate the voltage through energy storage in an inductor and filtering by a capacitor. This converter topology is extensively employed in solar photovoltaic systems for managing battery charging and maximum power point tracking operations [5].

3.1.2 Boost converter

In contrast, the boost converter is well-suited for increasing the input voltage, making it particularly useful in photovoltaic systems and fuel cells where the voltage level must be raised before connecting to an inverter or battery system. The boost topology can attain high output voltages, but it necessitates careful control to avert instability under light load conditions [8].

3.1.3 Buck-boost converter

The buck-boost converter is a versatile topology that enables both step-up and step-down voltage conversion while producing an inverted output voltage. This flexibility makes it beneficial in applications such as battery management systems and isolated DC sources, where unpredictable voltage variations are encountered. However, the simplicity of the buck-boost converter comes at the cost of high output ripple and increased component stress when operating extreme duty cycles (see **Figure 2**).

3.1.4 Cuk, SEPIC, and Zeta converters

These topologies provide either non-inverted or inverted output voltages while exhibiting enhanced ripple performance and continuous current at both the input and output. The Cuk converter utilizes capacitive energy transfer and is well-suited for applications requiring low input/output current ripple. The SEPIC converter offers a non-inverting output and is commonly employed in automotive and photovoltaic systems. Similarly, the Zeta converter, which shares a similar structure to the SEPIC, delivers a non-inverted output with improved dynamic response [9].

3.2 Isolated converters

3.2.1 Flyback converter

The flyback converter is one of the most straightforward isolated converter architectures, making it well-suited for low-power applications like off-grid photovoltaic lighting systems or low-voltage auxiliary supplies. This topology stores energy in a transformer during the switch-on period and then releases it during the switch-off period. While the flyback converter is a cost-effective solution, it exhibits high voltage stress and limited efficiency when operating at higher power levels [10].

3.2.2 Forward converter

The forward converter, building upon the flyback topology, employs transformer magnetizing reset techniques to facilitate more efficient energy transfer. This converter is well-suited for mid-power applications, such as standalone inverters or distributed generation systems. Its key advantages include reduced switch stress and enhanced energy transfer efficiency [11].

3.2.3 Push-pull converter

The push-pull converter employs two switches to drive a center-tapped transformer, allowing for bidirectional magnetization. This topology exhibits high efficiency and is well-suited for isolated systems that require symmetrical drive, such as small-scale wind power applications or backup power supplies.

3.2.4 Half-bridge and full-bridge converters

These topologies are prevalent in medium- and high-power renewable energy systems. The half-bridge converter exhibits good efficiency and reduces transformer stress. Conversely, the full-bridge converter provides higher power output with improved transformer utilization, rendering it suitable for grid-connected photovoltaic inverters or large-scale battery systems [12].

3.2.5 Multi-port and multi-input converters

Multi-port converters can interconnect multiple energy sources and/or loads through a single power conversion stage. These topologies are particularly advantageous in hybrid renewable energy systems that combine photovoltaic, wind, and energy storage technologies. Key benefits include flexible power flow management, dynamic prioritization of energy sources, and reduced component count. Bidirectional power conversion capabilities are essential in applications where batteries or supercapacitors are used to supply energy back to the load or electrical grid [13].

3.2.6 Switched-capacitor and switched-inductor converters

These topologies forgo the use of bulky magnetic components, instead leveraging capacitors or inductors to accomplish voltage conversion. Switched-capacitor converters are a compact option well-suited for low-power, integrated applications such as IoT and wearable renewable energy devices. In contrast, switched-inductor converters boast enhanced step-up capabilities and can be cascaded to attain ultrahigh gain—a crucial requirement for low-voltage photovoltaic or thermoelectric generator systems [14].

4. Advanced converter topologies for renewable energy systems

As renewable energy systems evolve to meet increasing demands for efficiency, scalability, and power density, traditional converter topologies are often insufficient. To address these challenges, advanced converter architectures have emerged, offering superior performance, reduced stress on components, and enhanced compatibility with next-generation semiconductors and control systems. This section explores interleaved converters, modular multilevel converters (MMCs), resonant and soft-switching converters, wide-bandgap-based converters, and hybrid topologies (see **Figure 3**).

4.1 Interleaved converters

Interleaving entails connecting multiple converter phases in parallel, with their switching times staggered. This technique effectively diminishes input and output current ripple, distributes thermal and electrical stress across components, and enables greater overall efficiency. Interleaved converters prove particularly valuable in high-current applications, such as electric vehicle charging and large-scale photovoltaic systems [15].

An interleaved boost converter has the capability to sustain lower input current ripple even under high-power conditions, thereby enhancing electromagnetic

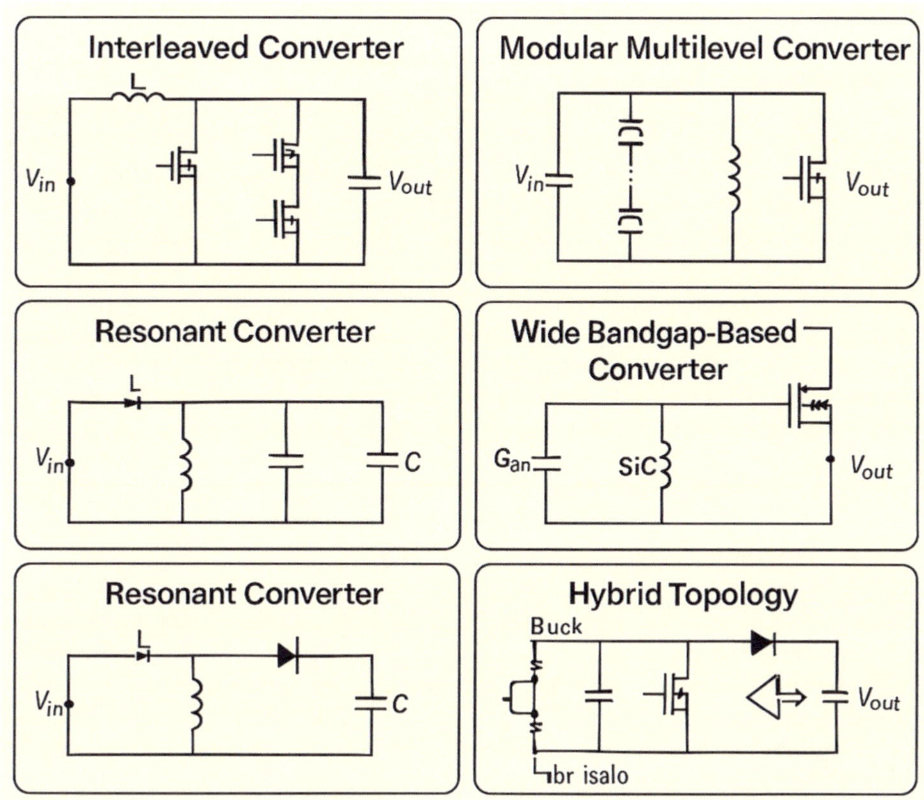

Figure 3.
Advanced converter topologies for renewable energy systems.

compatibility and reducing the size of the input capacitor. Although the control complexity escalates with the number of phases, it can be effectively managed through the utilization of digital controllers. Additionally, interleaved operation presents the advantage of improved transient response, a critical attribute in renewable energy systems that experience rapidly fluctuating inputs.

4.2 Modular multilevel converters (MMCs)

Originally developed for high-voltage direct current transmission, modular multilevel converters have been adopted in medium- and high-power renewable energy systems due to their scalability, high efficiency, and superior harmonic performance. MMCs consist of series-connected submodules, each capable of producing partial voltage steps. This modular architecture facilitates fine-grained voltage regulation and diminished harmonic distortion without the need for bulky filtering components [16].

Modular multilevel converters find extensive application in renewable energy systems, facilitating the grid integration of offshore wind farms, large-scale solar photovoltaic installations, and battery energy storage systems. The modular architecture of these converters also enables fault tolerance and the hot swapping of defective submodules, thereby enhancing overall system reliability. However, these topologies necessitate complex control algorithms and high-speed communication between their constituent submodules.

4.3 Resonant and soft-switching converters

Resonant converter topologies leverage inductors and capacitors to establish oscillating circuits, enabling either zero-voltage switching or zero-current switching, which substantially reduces switching losses. Common resonant converter variants include series resonant, parallel resonant, and LLC resonant configurations. These converter architectures are well-suited for applications necessitating high efficiency and low electromagnetic interference, such as high-frequency isolated power supplies and wireless energy transfer systems [17].

Soft-switching approaches expand beyond resonant operation, encompassing techniques such as active clamp and zero-transition methods that enhance efficiency without necessitating sizeable passive components. By mitigating switching losses and stress, these converter architectures enable higher switching frequencies, which in turn facilitate the use of smaller magnetic components and yield improved power density—objectives of paramount importance in portable and space-constrained renewable energy applications [18].

4.4 Wide-bandgap-based converters (GaN, SiC)

Wide-bandgap semiconductor materials, including gallium nitride and silicon carbide, have transformed the field of power electronics. These emerging semiconductor technologies provide faster switching capabilities, higher breakdown voltages, and improved thermal performance in comparison to traditional silicon-based devices. These advantageous characteristics enable power converters to operate at elevated frequencies and temperatures while maintaining reduced power losses [19].

Gallium nitride-based power converters are well-suited for low- to medium-power renewable energy applications, such as microinverters, portable photovoltaic systems, and on-board chargers for electric vehicles. Conversely, silicon carbide-based converters are more prevalent in high-power scenarios, including utility-scale solar farms and wind power systems. The integration of wide-bandgap semiconductor technologies, such as GaN and SiC, enhances the efficiency and power density of these converters, enabling the use of smaller passive components and improved thermal management capabilities. Gallium nitride devices have zero reverse recovery losses, very low gate drive losses, and low output charge compared to a silicon MOSFET, which makes GaN devices relevant for high-efficiency power converters [20].

4.5 Hybrid topologies

Hybrid converter architectures amalgamate attributes from diverse converter types to realize specialized performance objectives. For example, a buck-boost-flyback hybrid configuration can offer high voltage gain and galvanic isolation within a single stage. Additionally, converters that integrate interleaved and resonant structures are employed to strike a balance between efficiency, control complexity, and power density [21].

Hybrid converter architectures are becoming more prevalent in applications where a wide range of operating conditions are anticipated, such as hybrid renewable energy systems incorporating variable PV, wind, and battery sources. These converter topologies provide design flexibility, enhanced fault tolerance, and optimized energy management capabilities by dynamically adapting to changing load and source conditions (see **Figure 3**).

In summary, advanced DC-DC converter topologies play a pivotal role in enhancing the performance of renewable energy systems. Interleaved converters improve efficiency and reduce ripple, making them ideal for high-current applications like EVs and PV systems. Multilevel and modular multilevel converters enable high-voltage, medium-to-high power operation with low harmonic distortion, suitable for grid-connected and offshore systems. Resonant and soft-switching converters minimize switching losses and EMI, supporting compact, high-frequency designs. Bidirectional and isolated converters ensure safe, flexible power flow between sources, storage, and loads. Wide-bandgap semiconductor converters (GaN, SiC) allow faster switching with superior thermal performance. Z-source and quasi-Z-source converters offer single-stage buck-boost capability, reducing circuit complexity. Hybrid topologies integrate multiple converter features to handle dynamic and hybrid renewable conditions. Comparative evaluations of these designs highlight trade-offs in cost, complexity, and efficiency. These innovations collectively support the next generation of intelligent, modular, and high-performance renewable power systems.

5. Control techniques for DC-DC converters in renewable energy systems

Control strategies play a crucial role in shaping the performance, stability, and efficiency of DC-DC converters utilized in renewable energy systems. Advanced control approaches are indispensable in managing dynamic variations in source and load, optimizing power conversion, and ensuring reliable operation across a broad range of operating conditions. This section examines both traditional and contemporary control techniques, encompassing voltage and current mode control, hysteresis and sliding mode control (SMC), model predictive control (MPC), and digital control methods, as well as AI-based and adaptive control strategies (see **Figure 4**).

5.1 Voltage and current mode control

Voltage mode control is one of the most employed techniques in DC-DC converters. It operates by regulating the output voltage through modulation of the duty cycle in response to the voltage error signal. While votage mode control (VMC) offers simplicity and cost-effectiveness, it exhibits suboptimal dynamic performance under load transients and is more susceptible to variations in the input voltage.

Current mode control enhances the transient response and stability of DC-DC converters by incorporating an inner current loop alongside the outer voltage loop. This dual-loop architecture enables faster compensation of output voltage deviations and simplifies the design of the compensation networks. Current mode control (CMC) is particularly advantageous in renewable energy systems that experience rapid variations in power input, such as solar panels under fluctuating irradiance conditions [5].

5.2 Hysteresis control

Hysteresis control, also referred to as bang-bang control, operates by confining the output voltage or current within a predefined band around the reference value. The switching frequency varies depending on the system parameters and load conditions. This control approach provides a rapid dynamic response and inherent current-limiting capabilities, rendering it well-suited for applications such as PV maximum power point tracking and battery charging [22].

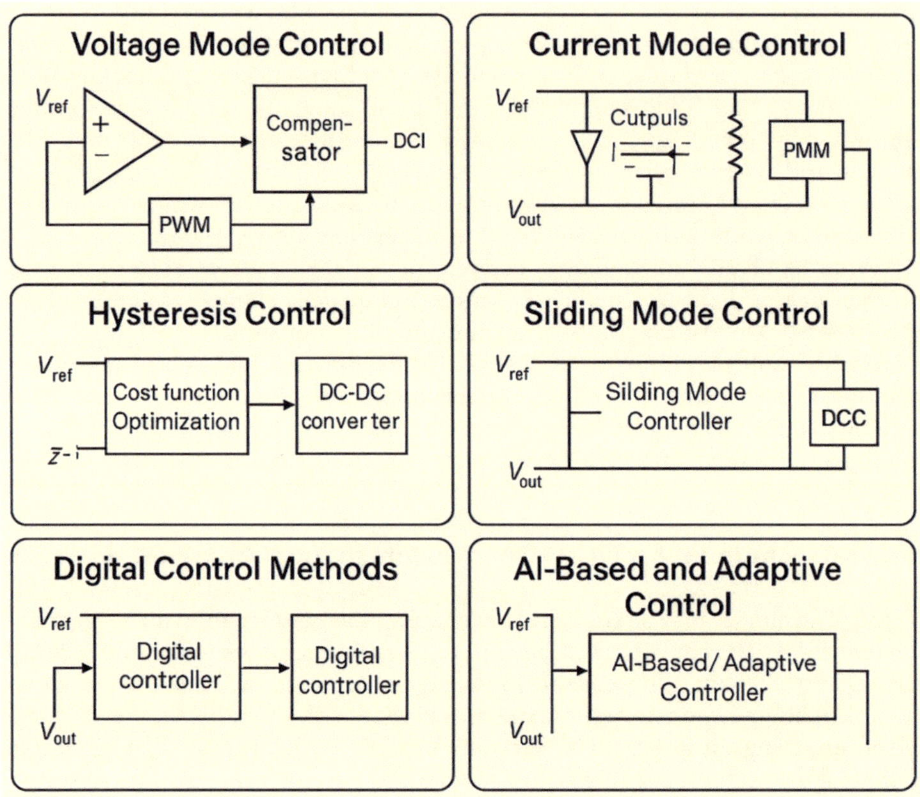

Figure 4.
Control techniques for DC-DC converters in renewable energy systems.

Despite its simplicity, hysteresis control exhibits limitations in electromagnetic interference performance due to its unpredictable switching frequency. To address this issue while maintaining the method's fast responsiveness, researchers are exploring techniques such as digital hysteresis control or adaptive hysteresis bands (see **Figure 4**).

5.3 Sliding mode control (SMC)

Sliding mode control is a nonlinear control approach renowned for its resilience against parameter variations and external perturbations. This technique compels the converter's behavior to adhere to a predefined sliding surface, enabling the system dynamics to be sustained on this surface despite model uncertainties or disturbances.

Sliding mode control is particularly well-suited for converters operating in harsh or unpredictable environments, such as standalone wind turbines or remote photovoltaic systems. However, this control technique can sometimes introduce chattering, which may lead to component stress. To address this issue, advanced implementations of SMC, such as second-order SMC or methods utilizing smooth sliding surfaces, have been developed to help mitigate the chattering problem [23].

5.4 Model predictive control (MPC)

Model predictive control leverages a dynamic model of the converter to forecast future behavior and determine the optimal control strategy over a finite timeframe. This approach factors in system constraints and performance objectives, enabling high-performance tracking and regulation capabilities. Model predictive control demonstrates particular efficacy in governing multivariable systems and nonlinear dynamics, rendering it well-suited for complex converter topologies, including multi-level and bidirectional converters utilized in energy storage systems, as well as hybrid renewable grids. This is attributed to its capacity to manage multiple objectives and constraints simultaneously. However, model predictive control is limited by its high computational complexity and the requirement for accurate system models [24].

5.5 Digital control methods

The emergence of high-performance microcontrollers and digital signal processors has led to the widespread adoption of digital control techniques in modern power electronics. Digital control architectures provide flexibility in algorithm implementation, enable real-time monitoring, and facilitate adaptive tuning capabilities. Furthermore, digital control schemes can seamlessly incorporate advanced control strategies, such as those based on artificial intelligence or neural networks, to enhance system performance and robustness.

Examples include discrete-time PI control, digital state feedback, and delay compensation techniques. Digital control also enables seamless integration with communication protocols and supervisory control in smart grid or IoT-enabled renewable systems [25].

5.6 AI-based and adaptive control algorithms

Emerging artificial intelligence and adaptive control methodologies are being increasingly integrated into DC-DC converter systems to enhance performance under dynamic and uncertain operating conditions. Techniques such as neural networks, fuzzy logic controllers, and genetic algorithms possess the capability to learn system behavior and dynamically adjust control parameters in real time (see **Figure 4**).

These techniques are particularly advantageous in complex renewable energy systems where load profiles, environmental conditions, or grid demands exhibit rapid variations. For instance, an AI-based maximum power point tracking controller can outperform traditional perturb-and-observe methods under fluctuating irradiance conditions [26].

Adaptive control techniques dynamically adjust controller parameters based on real-time system identification. This approach is particularly effective in systems with unknown or time-varying parameters, such as aging batteries or partially shaded photovoltaic arrays [27, 28].

6. Integration and applications

6.1 Integration with renewable energy sources

DC-DC converters play a crucial role in renewable energy systems, serving as the interface between variable and often intermittent sources, such as solar photovoltaics, wind turbines, and fuel cells, and the load or energy storage units. The integration of these converters into diverse energy systems necessitates consideration of the unique electrical characteristics and operational dynamics of each source. This section examines how DC-DC converters are tailored for use in various renewable energy applications, including solar PV systems, wind energy systems, fuel cells, hybrid energy systems, and energy storage devices like batteries and supercapacitors. Power electronics are very important for converting electrical energy from one form to another while still maintaining high conversion efficiency for solar PV-powered systems. **Figure 5** highlights the role of DC-DC converters in interfacing with solar PV systems, wind turbines, fuel cells, hybrid sources, and energy storage units. Each converter topology is tailored to manage unique source characteristics, ensure efficient power regulation, support MPPT algorithms, and enable seamless energy flow within hybrid and storage-integrated systems.

6.2 Solar PV systems

Solar PV systems experience significant variations in the output voltage and current of the panels due to changes in solar irradiance and temperature. To address this, DC-DC converters are employed to regulate the output, track the maximum power point, and ensure compatibility with downstream components like inverters or storage devices.

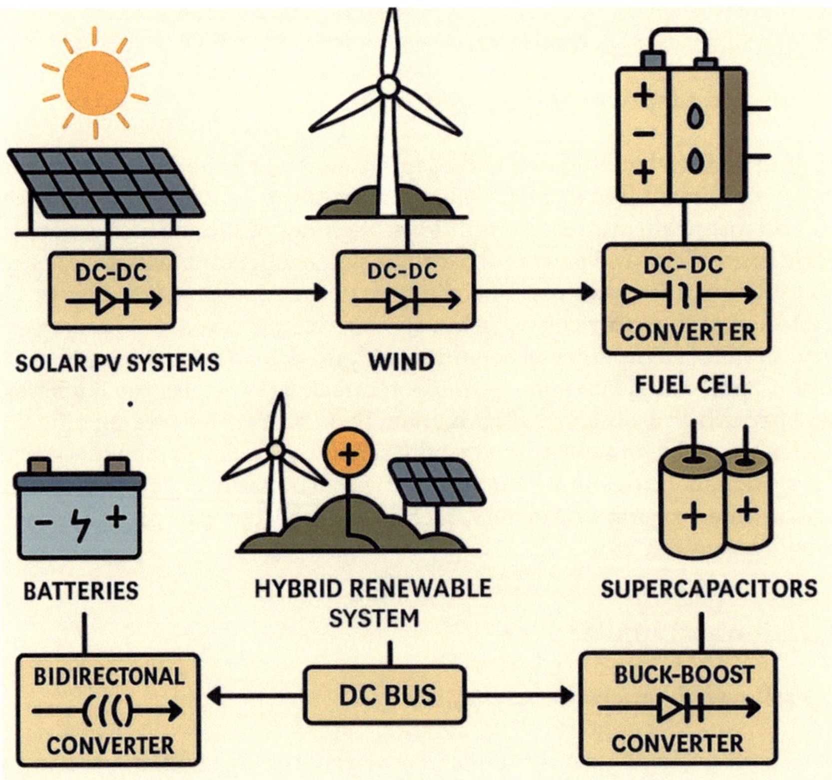

Figure 5.
Integration of DC-DC converters with renewable energy sources.

Commonly used converter topologies include buck, boost, and buck-boost configurations, selected based on the voltage requirements of the solar panels and the load. Advanced maximum power point tracking algorithms, such as perturb and observe, incremental conductance, or AI-based methods, are implemented in the converter's control scheme to continuously optimize power extraction from the solar panels [29].

Additionally, interleaved and resonant converter configurations are seeing growing adoption in microinverter designs, as they provide enhanced efficiency and improved thermal management capabilities that are advantageous for high-density photovoltaic installations (see **Figure 5**).

6.3 Wind energy systems

Unlike PV systems, wind turbines produce alternating current electricity, which is commonly converted to direct current before further processing. The fluctuations in wind speed result in varying voltage and frequency at the generator's output. To address this, a DC-DC converter is employed after the rectification stage to stabilize the DC link voltage and implement maximum power point tracking algorithms based on tip speed ratio or power signal feedback [30].

For small-scale wind energy applications, boost or buck-boost converter topologies are commonly employed due to their inherent simplicity. In contrast, larger wind energy systems utilizing permanent magnet synchronous generators often benefit from the enhanced efficiency and controllability offered by interleaved or soft-switching converter configurations.

6.4 Fuel cell systems

Fuel cells generate DC voltage that is relatively stable but low in magnitude and strongly influenced by the load. Employing a DC-DC converter is crucial to increase the voltage to a practical level while addressing the slow dynamic response and internal resistance inherent to the fuel cell stack.

High-gain converter topologies, such as cascade boost, interleaved boost, and coupled-inductor configurations, are commonly employed in fuel cell systems. Furthermore, specialized control strategies are implemented to mitigate rapid current variations that could potentially harm the fuel cell stack. In this context, current mode control and predictive control approaches are favored due to their ability to avoid such undesirable current changes [31].

The integration of DC-DC converters in fuel cell-powered systems, such as in electric vehicles or backup power units, also allows seamless operation with battery packs, which buffer the slow transient response of the fuel cells (see **Figure 5**).

6.5 Hybrid renewable systems (PV + wind, etc.)

Hybrid renewable energy systems integrate multiple generation sources to enhance reliability and energy supply. Due to disparities in voltage levels and operational characteristics, individual DC-DC converters may be necessary for each source. These converters then feed into a shared DC bus or a multi-input converter configuration.

Employing multi-port converter topologies can diminish the number of power conversion stages and enhance overall system efficiency by dynamically managing energy flows between sources, loads, and storage systems [32]. The DC-DC converters integrated with hybrid renewable systems must support bidirectional power flow capabilities, particularly when connected to energy storage units. Furthermore, they should implement intelligent control algorithms that prioritize energy sources based on their availability and the prevailing demand conditions.

6.6 Energy storage integration (batteries, supercapacitors)

Energy storage systems play a crucial role in maintaining the equilibrium between supply and demand in renewable energy installations. DC-DC converters are responsible for managing the charging and discharging processes, ensuring the protection of storage devices from issues such as overvoltage, overcurrent, and thermal stress.

Batteries typically employ bidirectional converter topologies, such as half-bridge or dual-active bridge configurations, to facilitate efficient energy exchange.

Supercapacitors, due to their rapid charge and discharge abilities, are coupled with buck-boost converters in applications requiring high power density and short-term energy storage [33].

Advanced control strategies enable dynamic energy management, peak shaving, and grid support in hybrid renewable systems with storage integration.

7. Applications of DC-DC converters in renewable energy systems

7.1 Grid-connected and off-grid applications

DC-DC converters are critical components in both grid-connected and off-grid renewable energy systems. They perform essential functions such as stabilizing voltage levels, enabling maximum power point tracking, and managing energy flow between sources, storage, and loads. This section examines the role and design considerations of DC-DC converters across various application areas, including grid-tied inverters, DC microgrids, off-grid systems, and emerging applications like vehicle-to-grid (V2G) and renewable-powered electric vehicle chargers (see **Figure 6**).

7.2 Role of DC-DC converters in grid-tied inverters

In grid-connected renewable energy systems, DC-DC converters are employed to condition the output of sources such as solar panels or fuel cells prior to interfacing with the grid through inverters. Their primary functions encompass voltage level adaptation, power quality enhancement, and the implementation of maximum power point tracking techniques. In photovoltaic systems, a DC-DC converter regulates the solar panel output to match the inverter's required input voltage while also executing real-time MPPT algorithms [34]. High-efficiency boost or buck-boost topologies are commonly used, particularly in string inverters or microinverters [35]. In some configurations, such as cascaded H-bridge multilevel inverters, the converters also help achieve voltage balancing across modules [36].

Advanced control methods are employed to ensure compliance with grid codes, such as low-voltage ride-through (LVRT), reactive power support, and harmonic control, making the converter a key enabler of smart grid integration.

Sophisticated control techniques are implemented to ensure conformance with grid codes, including low-voltage ride-through capabilities, reactive power support,

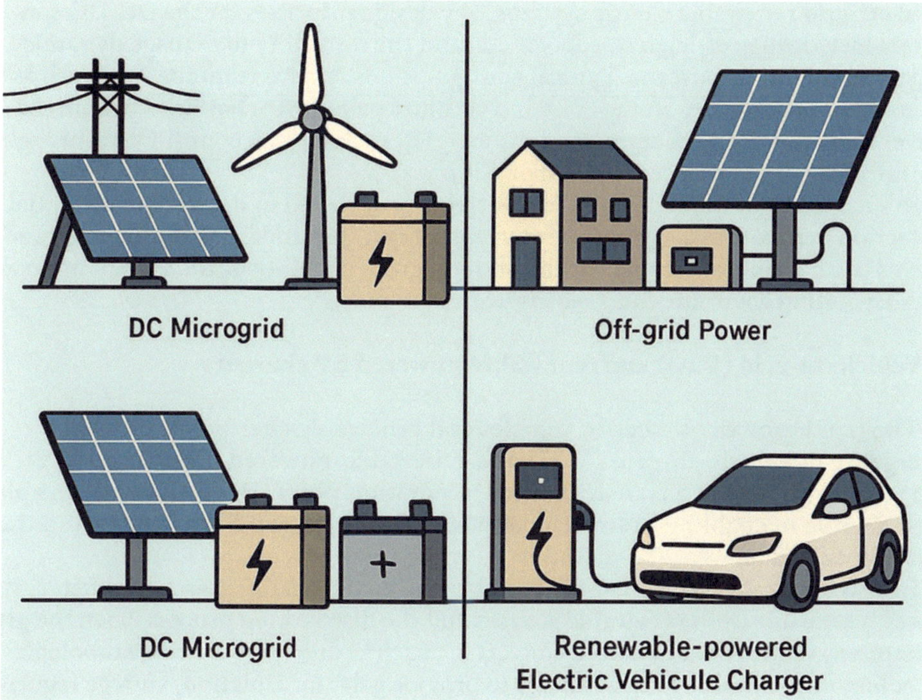

Figure 6.
Applications of DC-DC converters in renewable energy systems.

and harmonic control, positioning the converter as a crucial enabler for seamless integration with smart grid infrastructure (see **Figure 6**).

7.3 DC microgrids

DC microgrids are becoming increasingly popular in commercial buildings, data centers, military facilities, and rural electrification initiatives. These systems offer reduced conversion stages, enhanced efficiency, and simplified integration of DC-native power sources such as PV panels and batteries.

Within a DC microgrid, diverse sources, storage units, and loads are interconnected through a common DC bus. DC-DC converters regulate the power flow, balance the load, and maintain voltage stability across the grid. Bidirectional converters are employed for energy storage systems, while source-side converters ensure maximum power point tracking and isolation when required [37].

The converter's control architecture incorporates various power-sharing strategies, such as droop control, hierarchical control, and distributed energy management, enabling scalable and flexible system operation.

7.4 Off-grid power systems

In off-grid renewable energy systems—such as those utilized in remote regions, islands, or disaster relief scenarios—DC-DC converters are responsible for managing the energy flow between sources, storage units, and loads, operating independently of any grid connection.

In off-grid renewable energy systems, key design priorities for the DC-DC converters include high efficiency, robustness, and the capability to manage dynamic load and generation patterns. Typical configurations involve combinations such as PV coupled with battery storage or wind turbines paired with batteries, where the converters regulate the charging and discharging cycles, often coordinated through a centralized energy management system [38].

In such configurations, DC-DC converters are designed to incorporate essential protection mechanisms, such as overcurrent safeguards, undervoltage lockout, and thermal management systems, ensuring the overall reliability of the system in isolated and demanding environments (see **Figure 6**).

7.5 Vehicle-to-grid (V2G) and renewable-powered EV chargers

The rapid growth of electric vehicles and renewable energy has led to the emergence of vehicle-to-grid (V2G) and renewable-powered EV charging systems, which represent a transformative integration point. DC-DC converters play a crucial role in enabling bidirectional power flow between EV batteries and the grid or local loads.

In V2G applications, DC-DC converters manage the charging process when power is available from renewable sources and the discharging process when the grid demands it. Bidirectional isolated converters, such as dual-active-bridge topologies, are commonly used due to their ability to provide galvanic isolation, voltage regulation, and soft-switching capabilities [39].

Renewable-powered electric vehicle charging stations leverage DC-DC converters to directly connect with solar photovoltaic or wind power systems. By eliminating the need for multiple AC-DC and DC-AC conversion stages, these converters enhance overall efficiency and cost-effectiveness.

Additionally, intelligent control systems, often implemented through digital means, coordinate the charging process based on factors such as renewable energy generation availability, grid signals, or pricing algorithms. This creates a synergistic relationship between transportation electrification and sustainable energy sources.

8. Design challenges and considerations for DC-DC converters in renewable energy systems

8.1 Design challenges and considerations

The design and deployment of DC-DC converters in renewable energy systems are influenced by a range of technical and economic challenges. As power conversion requirements become increasingly stringent, with higher demands for efficiency, power density, and reliability, engineers must address critical design considerations. This section examines five key aspects: thermal management, electromagnetic interference, reliability and fault tolerance, performance under partial-load conditions, and constraints related to size, weight, and cost (see **Figure 7**).

8.2 Thermal management

Managing the heat generated during power conversion is a critical challenge in the design of DC-DC converters. Power losses, predominantly from conduction and

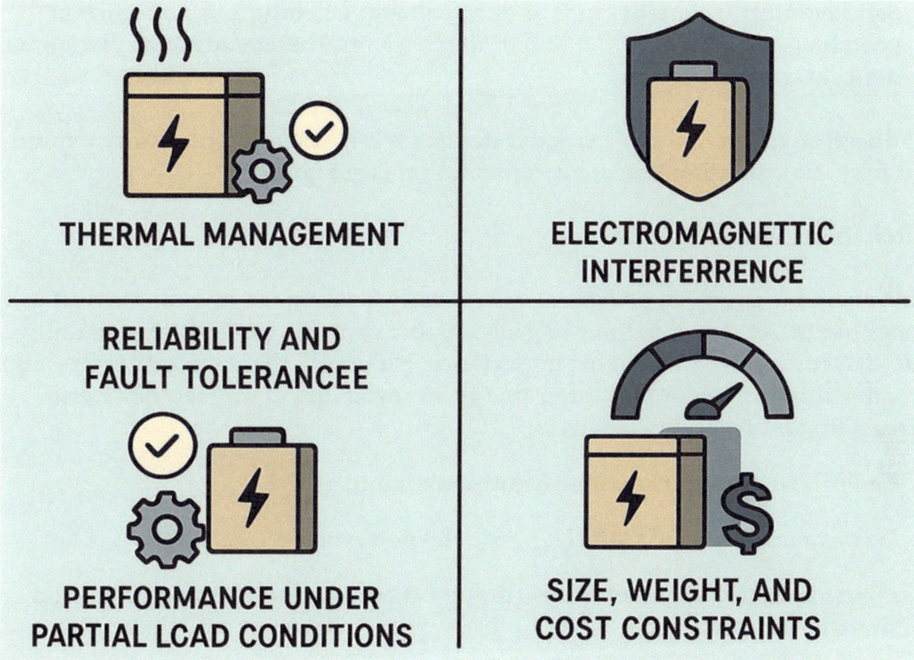

Figure 7.
Design challenges and considerations for DC-DC converters in RC-DC renewable energy systems.

switching, lead to substantial thermal accumulation, particularly in high-power and high-frequency applications common in renewable energy systems.

Effective thermal management is essential for ensuring component durability and system reliability. Strategies include the use of heat sinks, forced-air or liquid cooling, thermal *vias* in printed circuit boards, and advanced packaging materials such as aluminum nitride and metal-core printed circuit boards (PCBs) [40].

Designers must also optimize the layout and switching frequency to mitigate thermal hotspots. Wide-bandgap semiconductors like GaN and SiC provide superior thermal performance, enabling more efficient and compact converter designs by lowering switching losses [41].

8.3 Electromagnetic interference (EMI)

High-frequency switching in contemporary converters can produce substantial electromagnetic interference, which may disrupt nearby electronic devices and compromise overall system performance. EMI can manifest through both radiated and conducted paths, posing particular challenges in scenarios where power converters are collocated with sensitive control or communication equipment.

To mitigate EMI, designers employ several strategies:

- *Filtering*: Input and output EMI filters (LC or π-type) are commonly used.
- *Shielding*: Metal enclosures and ground planes reduce radiated noise.
- *PCB layout*: Careful trace routing, minimized loop areas, and ground isolation help reduce noise coupling.

- *Soft switching*: Techniques such as zero-voltage switching and zero-current switching can mitigate dv/dt and di/dt transients, thereby reducing electromagnetic interference emissions [42].

Adherence to relevant international standards is frequently a necessary requirement for successful commercial implementation (see **Figure 7**).

8.4 Reliability and fault tolerance

In renewable energy applications, particularly in remote or mission-critical settings like off-grid installations or grid-support systems, ensuring the reliability of DC-DC converters is of utmost importance. Failures in these systems can result in costly downtime, a loss of harvested energy, or instability in the electricity grid.

Designing for reliability involves:

- *Component derating* to operate below maximum ratings.
- *Redundancy* in critical paths (e.g., parallel power stages).
- *Protective features*, such as overcurrent, overvoltage, thermal shutdown, and short-circuit protection.
- *Health monitoring* and predictive maintenance using sensors and smart diagnostics.

Emerging solutions encompass fault-tolerant converter topologies that enable continued operation during partial component failures, as well as digital control systems that dynamically adapt to fault conditions in real time [43].

8.5 Efficiency under partial load conditions

Renewable energy sources frequently operate under fluctuating environmental conditions. For example, solar panels may generate reduced power on overcast days, and wind turbines vary with wind speed. Consequently, converters must sustain high efficiency across a broad load range, not solely at peak load.

This poses a particular challenge, as switching and core losses become more prominent compared to load power at light loads. Some solutions include [44]:

- *Burst mode operation*, which reduces switching frequency during light loads.
- *Dynamic biasing*, adjusting gate drive strength to match load.
- *Phase shedding* in interleaved converters, disabling phases to minimize idle losses.

Utilizing efficiency mapping and simulation techniques facilitates the assessment of performance across the anticipated operating range.

8.6 Size, weight, and cost constraints

For applications such as residential solar inverters, electric vehicles, and portable renewable devices, size and weight are crucial considerations. Simultaneously, cost must remain competitive to enable widespread adoption.

To achieve compact and cost-effective designs:

- *High-frequency operation*, which allows for smaller magnetics.
- *Integrated magnetics*, combining inductors and transformers.
- *Monolithic power modules*, using advanced packaging to reduce parasitics and footprint.
- *Economies of scale* and component standardization to lower bill-of-materials (BOM) costs.

Compromises frequently arise between size, thermal performance, and electromagnetic interference compliance, requiring a comprehensive and integrated design approach.

9. Recent trends and innovations in DC-DC converter technology for renewable energy systems

9.1 Recent trends and innovations

The field of DC-DC converter technology is undergoing rapid transformation to address the intricate requirements of contemporary renewable energy systems. Advancements in materials, control methodologies, and digital integration are catalyzing innovations that elevate the efficiency, flexibility, and intelligence of power conversion units. This section emphasizes some of the most impactful recent trends, encompassing smart converters, high-frequency designs, plug-and-play modules, reconfigurable architectures, and integration with the Internet of Things and cloud-based monitoring platforms (see **Figure 8**).

9.2 Smart DC-DC converters

Smart DC-DC converters leverage sophisticated sensing, control, and communication functionalities to dynamically optimize their operation. In contrast to traditional converters with fixed control schemes, these intelligent units can adapt to evolving environmental conditions, load demands, and grid requirements through real-time data analytics and artificial intelligence algorithms (see **Figure 5**).

Typically, smart converters employ digital controllers based on microcontrollers, DSPs, or FPGAs, which enable features such as self-optimization, fault detection, and predictive maintenance. Their ability to communicate over standardized protocols allows them to participate in energy management systems, positioning them as key enablers of modern smart grids and microgrids [45].

9.3 High-frequency and high-density converters

The drive for more compact, lightweight, and efficient power converters has spurred the embrace of high-frequency operation, enabled by the utilization of wide-bandgap semiconductors such as gallium nitride and silicon carbide. Operating at frequencies in the MHz range significantly diminishes the size of passive components

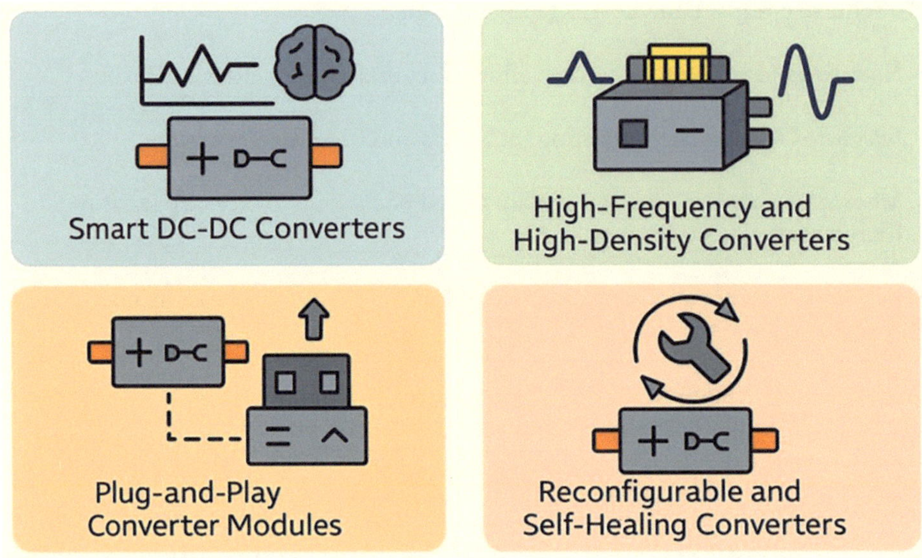

Figure 8.
Recent trends and innovations in DC-DC converter technology for renewable energy systems.

like inductors and capacitors, leading to high-density converter designs that are highly space-efficient.

While high-frequency switching enhances transient response and power density, it also introduces challenges in electromagnetic interference and thermal dissipation, necessitating the deployment of advanced materials and design techniques. These innovations are pivotal for applications where space and weight are at a premium, such as electric vehicles, aerospace, and distributed renewable energy generation [46].

9.4 Plug-and-play converter modules

Modular designs are an emerging trend in power electronics, driven by the need for scalable, sustainable, and readily integrable solutions. Plug-and-play DC-DC converter modules are standardized components that can be rapidly deployed and interconnected to construct larger systems with minimal engineering overhead.

These modular units often incorporate built-in safeguards, communication interfaces, and configurable control algorithms, enabling autonomous operation or coordinated functionality with other modules [47]. This modular design approach streamlines installation, facilitates system upgrades, and bolsters fault tolerance through simplified component replacement or parallel operation (see **Figure 8**).

9.5 Reconfigurable and self-healing converters

To enhance resilience and operational flexibility, reconfigurable and self-healing converter architectures are emerging as a prominent trend. These designs incorporate adaptable circuit topologies and control algorithms that can dynamically modify the power flow or operating mode in response to fault conditions or changing system requirements.

For instance, a converter may be able to bypass malfunctioning components, redistribute the load among healthy modules, or switch between isolated and non-isolated operation modes to sustain power delivery [48]. Self-healing capabilities in power converters help mitigate downtime and lower maintenance expenses, which is particularly crucial for remote or mission-critical renewable energy installations.

9.6 Integration with IoT and cloud-based monitoring

The incorporation of DC-DC converters with IoT platforms and cloud-based monitoring systems represents a significant advancement in the digitalization of renewable energy infrastructure. Sensors embedded within the converters collect real-time data on voltage, current, temperature, and efficiency metrics, which is then transmitted *via* wireless or wired networks to cloud-based servers. Cloud analytics leverage big data and machine learning techniques to provide insights into system health, predict potential failures, optimize energy management, and enable remote control capabilities. This interconnectivity supports the deployment of advanced grid services, such as demand response, virtual power plants, and enhanced cybersecurity, thereby bridging the realm of power electronics with the broader energy internet ecosystem [49].

10. Future trends in DC-DC converters for renewable energy systems

10.1 Future directions

The development of DC-DC converters in renewable energy systems is set to accelerate, propelled by advancements in artificial intelligence, smart grid architectures, emerging conversion technologies, and the integration of advanced energy storage. This chapter concludes by exploring the promising future directions that will shape the design, operation, and influence of DC-DC converters in sustainable power systems.

10.2 AI and machine learning in converter design

Artificial intelligence and machine learning have transformed numerous engineering domains, and power electronics is no exception. Integrating AI into the design and control of DC-DC converters holds immense potential for enhancing performance beyond traditional methods.

Machine learning algorithms can analyze extensive real-time sensor data to forecast system behaviors, optimize control parameters, and identify anomalies or faults before they escalate. For example, reinforcement learning techniques allow converters to dynamically adjust switching patterns and operating points, improving efficiency and reliability under varying load and environmental conditions [50].

Additionally, AI-assisted design automation allows engineers to explore extensive design spaces more effectively, striking a balance among efficiency, size, cost, and thermal performance. This data-driven approach expedites innovation and produces converters customized for specific renewable energy applications.

10.3 Role in future smart grids and IoT-enabled systems

In future smart grids and IoT-enabled infrastructures, DC-DC converters will assume a pivotal role as the energy systems become increasingly decentralized and interconnected. These converters will not only handle power conversion but also function as intelligent nodes capable of communication, control, and coordination across distributed energy resources.

Within such networks, the converters will enable bidirectional power flow, support demand response, and facilitate real-time grid services such as frequency regulation and voltage support. Moreover, the integration of IoT technologies will allow the converters to participate in predictive maintenance, dynamic load balancing, and cybersecurity frameworks [51].

These advancements will serve as the foundation for the envisioned highly responsive, resilient, and sustainable power grid, where power electronics devices are seamlessly integrated into digital energy ecosystems (see **Figure 9**).

10.4 Ultrafast and ultraefficient conversion technologies

The pursuit of ultrahigh-speed switching and exceptionally efficient power conversion is fueling advancements in device physics and converter architectures. Innovations in wide-bandgap semiconductors, ultrafast gate drivers, and novel circuit topologies enable switching frequencies reaching the tens of megahertz range, significantly reducing the size and losses of passive components.

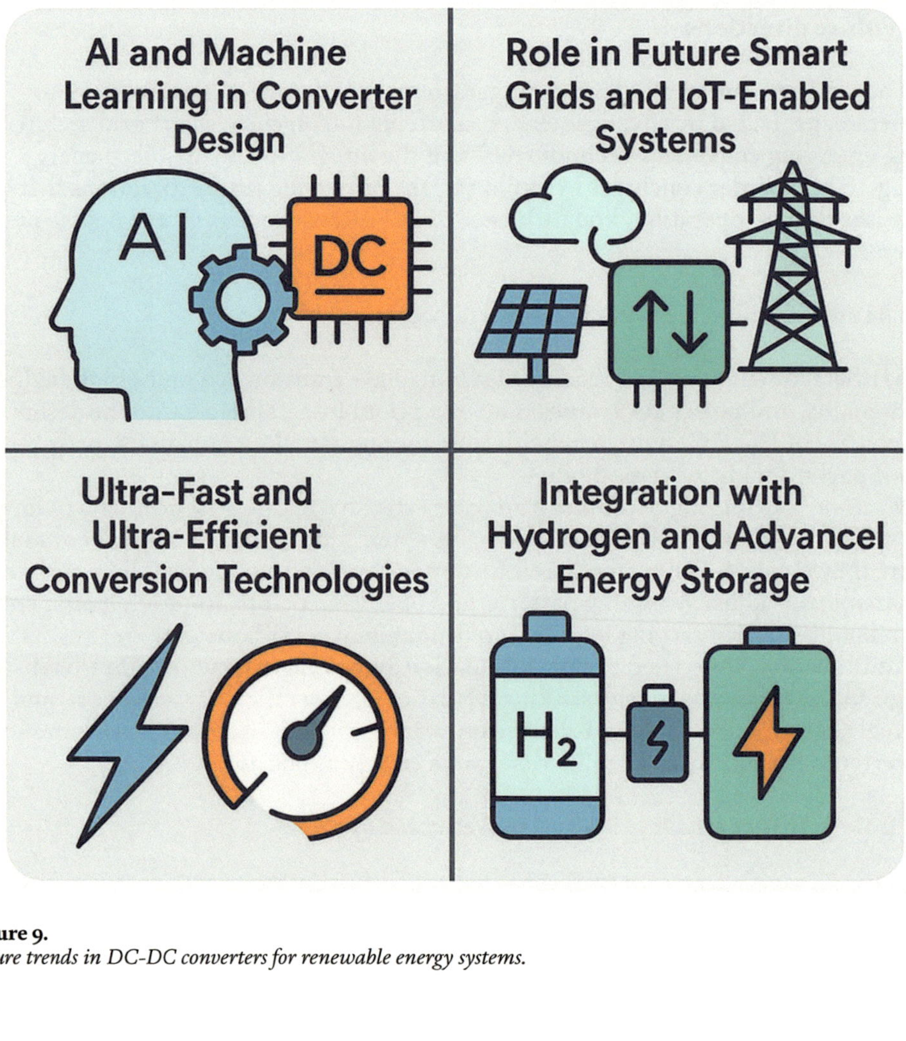

Figure 9.
Future trends in DC-DC converters for renewable energy systems.

DOI: http://dx.doi.org/10.5772/intechopen.1012305

Future converters are anticipated to incorporate resonant and soft-switching techniques that reduce switching losses and electromagnetic interference, enabling near-optimal efficiency even at partial load. Emerging concepts such as quantum tunneling devices and nanoscale semiconductors are also being explored for the next generation of power electronics [52].

These technological advancements will facilitate the downsizing of renewable energy converters, enabling their incorporation into a wide range of applications, from portable electronics to large-scale grid infrastructures (see **Figure 9**).

10.5 Integration with hydrogen and advanced energy storage

Hydrogen technologies and advanced energy storage systems are essential components of future renewable energy systems. DC-DC converters will play a crucial role in integrating renewable sources with electrolysis, fuel cells, and emerging storage technologies such as solid-state batteries and supercapacitors. For hydrogen production, converters must efficiently manage power flow to electrolyzers, often under variable renewable input conditions. Likewise, fuel cell systems necessitate stable and regulated DC power, requiring converters with precise control and high reliability [53].

Integration of advanced energy storage systems necessitates converters capable of bidirectional power flow, rapid charge and discharge cycles, and extended operational lifetimes. Intelligent converter architectures coupled with AI-based control will optimize energy management across hybrid storage solutions, thereby enhancing grid stability and the penetration of renewable energy.

11. DC-DC converter applications in renewable energy: Case studies and implementations

11.1 Case studies and practical implementations

This chapter showcases selected case studies and practical implementations of DC-DC converters in renewable energy systems, bridging the gap between theory and real-world applications. These examples illustrate the effectiveness of diverse converter topologies, control strategies, and integration techniques, supported by simulation and experimental data. The insights gained highlight the best practices, challenges, and lessons learned in the deployment of power electronics in renewable energy scenarios (see **Figure 10**).

11.2 Real-world applications in renewable energy systems

DC-DC converters are ubiquitous in renewable energy applications, spanning from small-scale residential solar installations to large-scale utility-level wind farms. For example, in photovoltaic systems, boost converters are extensively utilized to step up the panel voltage to align with the requirements of battery storage or inverter inputs, thereby facilitating maximum power point tracking and optimizing energy conversion efficiency [54].

In wind energy systems, isolated converter topologies such as full-bridge or push-pull are frequently utilized to manage the variable outputs from turbines and provide galvanic isolation, which enhances safety and mitigates noise. Furthermore, in hybrid renewable energy systems integrating both solar and wind power, multi-input converters are employed to optimize the power flow from multiple sources, thereby maximizing resource utilization and improving overall system reliability [55].

Studies of off-grid microgrids demonstrate how DC microgrids relying on converter-interfaced renewable energy sources can expand energy access in remote regions, offering stable, controllable, and scalable power solutions [56].

11.3 Comparative analysis of converter topologies

Selecting the appropriate converter topology is crucial in balancing efficiency, cost, complexity, and performance demands. Comparative analyses have evaluated the performance of fundamental topologies—including buck, boost, and buck-boost—against more intricate isolated converters and multi-port architectures. For instance, buck converters exhibit simplicity and high efficiency in applications requiring voltage step-down but lack galvanic isolation. Boost converters excel when voltage step-up is necessary yet often experience high voltage stress on components. Conversely, buck-boost converters offer bidirectional power flow and wide input/output voltage ranges, advantageous in battery charging applications [5].

Isolated converter topologies, such as the full-bridge, offer galvanic isolation, enhancing safety and noise immunity. However, this advantage comes at the expense of increased circuit complexity and physical size. Conversely, multi-port converters are well-suited for hybrid renewable energy systems but necessitate sophisticated control algorithms to manage power flow from diverse sources [57].

Comparative analyses provide essential guidance to system designers, enabling them to select converter topologies optimized for specific renewable energy applications (see **Figure 10**).

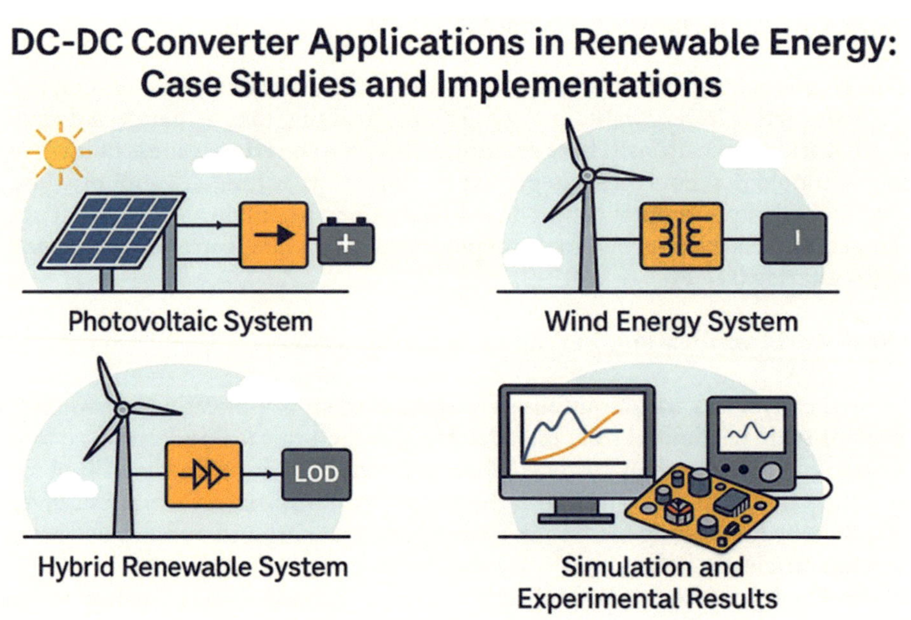

Figure 10.
DC-DC converter applications in renewable energy: Case studies and implementation.

11.4 Simulation and experimental results

Simulation tools such as MATLAB/Simulink, PLECS, and PSIM are widely employed to model the behavior of DC-DC converters under diverse operating conditions. These simulations enable designers to optimize component dimensions, control parameters, and thermal management strategies before physical implementation. For instance, simulating a solar PV boost converter illustrates how advanced control techniques, such as perturb and observe, can enhance the maximum power point tracking efficiency compared to incremental conductance methods [58]. Modeling an interleaved buck converter for wind turbine power conditioning demonstrates reduced input current ripple and enhanced transient performance compared to single-phase converter architectures [59].

Verifying simulation results and evaluating real-world performance require experimental validation. Laboratory prototypes of wide-bandgap-based converters demonstrate enhanced switching speeds and thermal characteristics, aligning with theoretical predictions. Moreover, field tests in grid-connected renewable energy installations validate the reliability and efficiency advantages of modular and reconfigurable converter designs [60].

The integration of simulation and experimental data offers a comprehensive framework for evaluating DC-DC converters, enabling their confident deployment in renewable energy systems.

12. DC-DC converters in renewable energy systems: Overview, challenges, and future directions

12.1 Conclusion

This concluding section provides a comprehensive overview of DC-DC converters in renewable energy systems, encompassing fundamental principles, topologies, control techniques, integration strategies, challenges, and emerging trends. Here, we synthesize the key insights gained throughout the chapter and offer recommendations to guide researchers and practitioners toward future advancements in this critical field.

12.2 Summary of key insights

DC-DC converters are indispensable components in renewable energy systems, enabling efficient, reliable, and flexible power conversion between sources, storage, and loads. The diversity of converter topologies—from basic buck and boost to complex isolated and multi-port configurations—allows tailored solutions to meet specific voltage, current, and isolation requirements. Advanced topologies such as interleaved and resonant converters further improve performance by reducing losses and enhancing dynamic response [5].

Control techniques play a critical role in optimizing converter operation. While voltage and current mode control remain foundational, emerging methods like model predictive control and AI-based adaptive algorithms offer enhanced precision and adaptability. The integration of converters with renewable sources—solar PV, wind, fuel cells—and hybrid systems necessitates robust power management and coordination with energy storage elements to ensure stable and maximized energy harvest [61].

Practical applications demonstrate the real-world relevance of these technologies, with case studies showcasing modular designs, smart converters, and digital integration supporting grid-tied and off-grid systems. Ongoing challenges such as thermal management, electromagnetic interference, and reliability under partial load highlight areas for continued innovation. Recent trends emphasize the role of wide-bandgap semiconductors, high-frequency operation, plug-and-play modularity, and IoT connectivity in advancing converter capabilities. Looking ahead, AI and machine learning are poised to transform converter design and control, enabling real-time optimization and predictive maintenance. The integration of converters into smart grids and IoT-enabled systems will facilitate more responsive and resilient energy networks. Furthermore, ultrafast switching technologies and the incorporation of hydrogen and advanced storage solutions will expand the operational horizons of DC-DC converters, addressing future energy demands [53].

12.3 Recommendations for researchers and practitioners

- *Embrace digital and AI technologies:* Researchers are encouraged to prioritize the development of AI-powered control algorithms and digital simulation models for DC-DC converters. These advanced tools can enable real-time adaptive control, predictive maintenance, and automated design, paving the way for more intelligent and autonomous power conversion systems.

- *Focus on wide-bandgap semiconductor integration:* The integration of wide-bandgap semiconductors, such as GaN and SiC, into DC-DC converters presents significant opportunities to enhance efficiency and power density. However, future research should focus on addressing the technical challenges associated with these materials, including thermal management, electromagnetic interference mitigation, and cost reduction, in order to facilitate their broader adoption and deployment.

- *Advance modular and reconfigurable architectures:* Practitioners should investigate modular converter architectures that facilitate scalability, maintainability, and system flexibility. Reconfigurable converters capable of adapting to evolving operating conditions or fault situations can bolster the reliability and resilience of renewable energy installations.

- *Optimize converter-source-storage integration:* Developing integrated energy management strategies that coordinate the operation of DC-DC converters with renewable energy sources and storage systems can enhance overall system efficiency and stability. Continued research into multi-input and hybrid converter topologies holds promise for further advancements in this area.

- *Prioritize cybersecurity and IoT compatibility:* As power converters become interconnected components within smart grids, ensuring robust cybersecurity measures and seamless IoT integration is crucial. Researchers should develop secure communication protocols and cloud-based monitoring platforms to safeguard and optimize renewable energy assets.

- *Address real-world challenges:* Ongoing research is essential to address practical challenges, including thermal management, electromagnetic interference mitigation, partial-load efficiency optimization, and reliability under adverse environmental conditions. Experimental validation and field testing should complement simulation studies to verify the efficacy of proposed designs and control strategies.

12.4 Final thoughts

DC-DC converters are essential components in the renewable energy transition, providing the power electronics infrastructure required to effectively harness, transform, and manage renewable electricity. The advancements in topology, control, and system integration outlined in this chapter demonstrate the vibrant and dynamic nature of this field.

By leveraging emerging technologies and addressing practical design challenges, researchers and industry professionals can develop next-generation converters that not only enhance energy conversion efficiency but also contribute to the development of smarter, more resilient, and sustainable power systems globally. Collaboration between academia, industry, and policymakers will be pivotal in realizing this vision and accelerating the widespread adoption of renewable energy solutions [51].

13. Conclusion

DC-DC converters are indispensable components in modern renewable energy systems, serving not only as voltage regulators but also as critical enablers of dynamic power management, energy optimization, and seamless system integration. Their evolution from simple converters to sophisticated, intelligent modules reflects the growing complexity and demands of sustainable energy technologies.

The diversity of converter topologies—ranging from basic buck-boost designs to advanced resonant, modular, and hybrid configurations—offers tailored solutions based on efficiency, scalability, isolation, and control needs. Innovations such as wide-bandgap semiconductors, soft-switching techniques, and modular architecture are pushing performance boundaries, enhancing the efficiency and adaptability of renewable systems.

Equally vital are advanced control strategies, which span from traditional voltage regulation to adaptive and AI-based approaches, each balancing complexity, cost, and performance. The integration of digital control enables sophisticated, real-time adjustments, reinforcing system stability and responsiveness.

DC-DC converters also face multifaceted design challenges, including thermal management, EMI mitigation, partial-load efficiency, and compact form factors. Addressing these challenges is essential to ensure reliability and scalability in both grid-connected and standalone applications such as DC microgrids, electric vehicles, and V2G infrastructures.

Furthermore, the role of converters in practical implementations—validated through case studies, simulations, and experimental data—confirms their pivotal contribution to real-world renewable energy deployments. As digital networking, advanced materials, and intelligent control continue to shape future energy ecosystems, DC-DC converters will remain at the forefront of enabling resilient, efficient, and intelligent power networks.

In conclusion, the ongoing innovation and strategic application of DC-DC converters are vital to the success of renewable energy integration, driving the transition toward a cleaner, decentralized, and more sustainable global energy future.

Author details

Mohammed Gronfula
College of Engineering, Alasala Colleges, Dammam, Saudi Arabia

*Address all correspondence to: mohammed.gronfula@alasala.edu.sa

References

[1] Patel H, Agarwal V. MPPT scheme for a PV-fed single-phase single-stage grid-connected inverter operating in CCM with only one current sensor. IEEE Transactions on Energy Conversion. 2009;**24**(1):256-263

[2] Kizilyalli IC et al. Wide Band-Gap Semiconductor Based Power Electronics for Energy Efficiency. Washington DC (United States): US Department of Energy (USDOE), Advanced Research Projects Agency-Energy (ARPA-E); 2018

[3] Khaligh A, Onar OC. Energy Harvesting: Solar, Wind, and Ocean Energy Conversion Systems. Boca Raton: CRC Press; 2017

[4] Wu TS, Husain I, Ehsani M. Switched-mode converters for high-power DC-DC applications. International Journal of Electronics. 1994;**77**(5):583-599

[5] Erickson RW, Maksimovic D. Fundamentals of Power Electronics. New York, NY: Springer Science and Business Media; 2007

[6] Ramachandran R, Nymand M. Evaluation of an un-regulated input isolated DC–DC converter using WBG devices. The Journal of Engineering. 2019;**2019**(17):3559-3562

[7] Suntio T, Hankaniemi M, Karppanen M. Analysing the dynamics of regulated converters. IEE Proceedings-Electric Power Applications. 2006;**153**(6):905-910

[8] Jain S, Agarwal V. A new algorithm for rapid tracking of approximate maximum power point in photovoltaic systems. IEEE Power Electronics Letters. 2004;**2**(1):16-19

[9] Fakhry M et al. Design and analysis of controlled SEPIC and LLC converters for photovoltaic systems. International Journal of Engineering Applied Sciences and Technology. 2020;**5**(8):54-63

[10] Mohan N, Undeland TM, Robbins WP. Power Electronics: Converters, Applications, and Design. 3rd ed. Hoboken, NJ: John Wiley & Sons; 2003

[11] Wu B, Narimani M. High-Power Converters and AC Drives. Hoboken, NJ: John Wiley & Sons; 2017

[12] Rodriguez J, Lai J-S, Peng FZ. Multilevel inverters: A survey of topologies, controls, and applications. IEEE Transactions on Industrial Electronics. 2002;**49**(4):724-738

[13] Zeng J et al. An isolated multiport DC–DC converter for simultaneous power management of multiple different renewable energy sources. IEEE Journal of Emerging and Selected Topics in Power Electronics. 2013;**2**(1):70-78

[14] Das M, Agarwal V. Design and analysis of a high-efficiency DC–DC converter with soft switching capability for renewable energy applications requiring high voltage gain. IEEE Transactions on Industrial Electronics. 2016;**63**(5):2936-2944

[15] Henn GA et al. Interleaved-boost converter with high voltage gain. IEEE Transactions on Power Electronics. 2010;**25**(11):2753-2761

[16] Marquardt R. Modular multilevel converter topologies with DC-short circuit current limitation. In: 8th International Conference on Power

Electronics-ECCE Asia. Jeju, Korea (South): IEEE; 2011. DOI: 10.1109/ICPE.2011.5944451

[17] Kazimierczuk MK, Czarkowski D. Resonant Power Converters. Hoboken, New Jersey, USA: John Wiley & Sons; 2012

[18] Gautam DS et al. A zero-voltage switching full-bridge DC--DC converter with capacitive output filter for plug-in hybrid electric vehicle battery charging. IEEE Transactions on Power Electronics. 2013;**28**(12):5728-5735

[19] Vechalapu K et al. Comparative evaluation of 15-kV SiC MOSFET and 15-kV SiC IGBT for medium-voltage converter under the same dv/dt conditions. IEEE Journal of Emerging and Selected Topics in Power Electronics. 2016;**5**(1):469-489

[20] Jagadeesh I, Indragandhi V. A review on isolated DC–DC converters used in renewable power generation applications. Artificial Intelligent Techniques for Electric and Hybrid Electric Vehicles. 2020;**1**:233-240

[21] Mondol MH et al. A new hybrid multilevel inverter topology with level shifted multicarrier PWM technique for harvesting renewable energy. IEEE Transactions on Industry Applications. 2022;**58**(2):2574-2585

[22] Kjaer SB, Pedersen JK, Blaabjerg F. A review of single-phase grid-connected inverters for photovoltaic modules. IEEE Transactions on Industry Applications. 2005;**41**(5):1292-1306

[23] Utkin V, Guldner J, Shi J. Sliding Mode Control in Electro-Mechanical Systems. Boca Raton: CRC press; 2017

[24] Geyer T, Papafotiou G, Morari M. Model predictive control in power electronics: A hybrid systems approach. In: Proceedings of the 44th IEEE Conference on Decision and Control and European Control Conference (CDC-ECC 2005) - "Melia Sevilla Hotel", Seville, Spain: IEEE; 2005

[25] Buso S, Mattavelli P. Digital Control in Power Electronics. Cham, Switzerland: Springer Nature; 2022

[26] Talaat M et al. Artificial intelligence applications for microgrids integration and management of hybrid renewable energy sources. Artificial Intelligence Review. 2023;**56**(9):10557-10611

[27] Tzempelikos A, Shen H. Comparative control strategies for roller shades with respect to daylighting and energy performance. Building and Environment. 2013;**67**:179-192

[28] Truntič M et al. Adaptive fuzzy-logic state controller for DC–DC step-down converter. Electrical Engineering. 2018;**100**(2):983-995

[29] Femia N et al. Optimization of perturb and observe maximum power point tracking method. IEEE Transactions on Power Electronics. 2005;**20**(4):963-973

[30] Kesraoui M, Korichi N, Belkadi A. Maximum power point tracker of wind energy conversion system. Renewable Energy. 2011;**36**(10):2655-2662

[31] Emadi A, Williamson SS, Khaligh A. Power electronics intensive solutions for advanced electric, hybrid electric, and fuel cell vehicular power systems. IEEE Transactions on Power Electronics. 2006;**21**(3):567-577

[32] Zhao C, Round SD, Kolar JW. An isolated three-port bidirectional DC-DC converter with decoupled power flow management. IEEE

Transactions on Power Electronics. 2008;**23**(5):2443-2453

[33] Jaguemont J, Boulon L, Dubé Y. A comprehensive review of lithium-ion batteries used in hybrid and electric vehicles at cold temperatures. Applied Energy. 2016;**164**:99-114

[34] Opiyo N. Power electronics for PV-based communal grids. Smart Grid and Renewable Energy. 2016;**7**:67-82

[35] Duong T-D et al. Topology review of three-phase two-level transformerless photovoltaic inverters for common-mode voltage reduction. Energies. 2022;**15**(9):3106

[36] Teodorescu R, Liserre M, Rodriguez P. Grid Converters for Photovoltaic and Wind Power Systems. Hoboken, New Jersey, USA: John Wiley & Sons; 2011

[37] Jian Z et al. A review of control strategies for DC micro-grid. In: 2013 Fourth International Conference on Intelligent Control and Information Processing (ICICIP). Beijing, China: IEEE; 2013. DOI: 10.1109/ICICIP.2013.6568157

[38] Kanase-Patil A, Saini R, Sharma M. Integrated renewable energy systems for off grid rural electrification of remote area. Renewable Energy. 2010;**35**(6):1342-1349

[39] Yilmaz M, Krein PT. Review of battery charger topologies, charging power levels, and infrastructure for plug-in electric and hybrid vehicles. IEEE Transactions on Power Electronics. 2012;**28**(5):2151-2169

[40] Muhammad H. Power Electronics: Devices, Circuits, and Applications. Boston, Massachusetts, USA: Pearson; 2014

[41] Flack TJ, Pushpakaran BN, Bayne SB. GaN technology for power electronic applications: A review. Journal of Electronic Materials. 2016;**45**:2673-2682

[42] Maksimovic D et al. Modeling and simulation of power electronic converters. Proceedings of the IEEE. 2002;**89**(6):898-912

[43] Musallam M et al. Mission profile-based reliability design and real-time life consumption estimation in power electronics. IEEE Transactions on Power Electronics. 2014;**30**(5):2601-2613

[44] Abd El-Aal AEMM et al. Modeling and Optimizing the Size of the Power Conditioning Unit for Photovoltaic Systems. New York, USA: American Society of Mechanical Engineers (ASME); 2006

[45] Bughneda A et al. Resonant power converters for renewable energy applications: A comprehensive review. Frontiers in Energy Research. 2022;**10**:846067

[46] Millan J et al. A survey of wide bandgap power semiconductor devices. IEEE Transactions on Power Electronics. 2013;**29**(5):2155-2163

[47] Chakraborty S, Kramer B, Kroposki B. A review of power electronics interfaces for distributed energy systems towards achieving low-cost modular design. Renewable and Sustainable Energy Reviews. 2009;**13**(9):2323-2335

[48] Lakshika KH et al. Reconfigurable solar photovoltaic systems: A review. Heliyon. 18 Nov 2020;6(11):e05530. DOI: 10.1016/j.heliyon.2020.e05530. eCollection 2020 Nov

[49] Hashmi SA, Ali CF, Zafar S. Internet of things and cloud computing-based

energy management system for demand side management in smart grid. International Journal of Energy Research. 2021;**45**(1):1007-1022

[50] Duan J et al. Reinforcement-learning-based optimal control of hybrid energy storage systems in hybrid AC–DC microgrids. IEEE Transactions on Industrial Informatics. 2019;**15**(9):5355-5364

[51] Erol-Kantarci M, Mouftah HT. Energy-efficient information and communication infrastructures in the smart grid: A survey on interactions and open issues. IEEE Communications Surveys & Tutorials. 2014;**17**(1):179-197

[52] Azizi M, Van Oorschot J, Huiskamp T. Ultrafast switching of SiC MOSFETs for high-voltage pulsed-power circuits. IEEE Transactions on Plasma Science. 2020;**48**(12):4262-4272

[53] Rego G et al. A review of hydrogen production methods and power electronics converter topologies for green hydrogen applications. Energies. 2024;**17**(22):5579

[54] Xu Q et al. Review on advanced control technologies for bidirectional DC/DC converters in DC microgrids. IEEE Journal of Emerging and Selected Topics in Power Electronics. 2020;**9**(2):1205-1221

[55] Kouro S et al. Recent advances and industrial applications of multilevel converters. IEEE Transactions on Industrial Electronics. 2010;**57**(8):2553-2580

[56] Katiraei F, Iravani MR, Lehn PW. Micro-grid autonomous operation during and subsequent to islanding process. IEEE Transactions on Power Delivery. 2005;**20**(1):248-257

[57] Chandrasekhar B, Lakshminarayanan S, Kumar S. Multi input DC-DC converters for hybrid renewable energy systems-an overview. Power Research-A Journal of CPRI. 2015:155-160. Available from: https://cprijournal.in/index.php/pr/issue/view/26

[58] Eseosa O, Kingsley I. Comparative study of MPPT techniques for photovoltaic systems. Saudi Journal of Engineering and Technology. 2020;**5**(02):38-48

[59] Lin J et al. Accurate loop gain modeling of digitally controlled buck converters. IEEE Transactions on Industrial Electronics. 2021;**69**(1):725-739

[60] Gurpinar E. Wide-Bandgap Semiconductor Based Power Converters for Renewable Energy Systems. Nottingham, United Kingdom: University of Nottingham; 2017

[61] Bayoumi EH. Power electronics in renewable energy smart grid: A review. International Journal of Industrial Electronics and Drives. 2015;**2**(1):43-61